PUBLISHED BY
CASACANTARILLA PUBLICATIONS
WWW.CASACANTARILLA.ORG.UK

CASA CANTARILLA IS AN ASSOCIATION
OF TEACHERS IN THE ARTS AND SCIENCES
OPERATING IN LANZAROTE
(CODIGA DE IDENTIFICACION G35750595)

Death by Eating

The evolution of human food

David McFarland

Contents

In which it is explained that human food is evolving as part of the human extended phenotype, the bodily expression of our genetic makeup. Over the past 15000 years our bodies have not changed much, but what we put into them has changed a lot. What we put into them is the result of our own behaviour, our own attitudes and our own desires. It is part of our evolution.

Hunger may seem to be a unitary phenomenon, but it is not. Hunger for water, carbohydrate, protein, salt, and many other specific factors strongly influences the way animals, including humans, behave when hungry. In this chapter the reader is introduced to the idea that there are many aspects of foraging and feeding, and many causes of death by eating. Moreover, the reader will figure in this book. In each chapter the reader will experience a death by eating. In each chapter the reader becomes another 'historical' person. Altogether the reader will have nine lives, and each life will end in some kind of death by eating.

CHAP 2   *THE EVOLUTION OF FEEDING STRATEGIES*

In this chapter the reader is introduced to evolutionary strategies, particularly those to do with feeding. As a result of the complex evolutionary interactions between plants and animals, humans have both enemies and friends when it comes to finding food.

CHAP 3   *THE DANGERS OF FORAGING*

Foraging for food requires the efficient expenditure of energy and time. Our remote ancestors were adept foragers as a result of thousands of years of evolution. But they still had to cope with rivals, enemies and dangerous plants.

CHAP 4   *GROWING YOUR OWN*

Those peoples that started to grow their own food and domesticate plants and animals underwent a drastic change in lifestyle. Their activities changed their environment in an irreversible manner, and they were no longer evolutionarily attuned to their diet. Consequently, their expectancy of life declined.

CHAP 5   *THE RISKS AND BENEFITS OF EXCHANGE*

Primitive agriculture leads to domestication of plants and animals, and to more complex economic organization. Except when individuals consume the fruits of their own labours, the products of human labour are distributed by means of exchange, the practice of giving and receiving valuable objects and services. Those individuals responsible for distributing food acquire a special (political) status. As a result, many people were no longer in control of their food supplies, and their choices became limited, and their dietary flexibility lost.

CHAP 6   *THE POWER OF ADVERTISING*

Advertising was evolved by plants and animals as a means of communication. It was developed as a means of persuasion, and even manipulation. In the human context, advertisers aim to promote one type of food at the expense of another, and to persuade people to do what they might not otherwise do. As a result, there is a distorting effect upon the human diet that is not always beneficial.

CHAP 7   *FOOD AND THE FAMILY*

The dietary preferences of grown individuals depend upon a host of factors, including their genetic makeup, their upbringing, and their subsequent lifestyle. Eating habits established in childhood are hard to change, and for working families, convenient foods are hard to resist. Food producers, in catering for (and sometimes manipulating) such situations, may preserve and 'enhance' food in ways that are not beneficial for the body, often resulting in the 'malnutrition of affluence.'

CHAP 8   *FOOD AND STATUS*

An important aspect of human food choice is peer pressure, the result of one's perceived social status. Such perceptions can have a disruptive influence upon diet. This is particularly true of people who worry about their perceived body image. The social aspect of food is often manipulated by food marketers, because there is money to be made out of changes in food 'fashion'.

CHAP 9   *FOOD PANIC*

Before a person puts food into their mouth, they have a belief about the food. Such beliefs are sometimes well founded, but they are often over-influenced by advertising, social pressures and doctrinaire assumptions. In evolutionary terms there have always been predators and parasites waiting to exploit such situations, and we see these in the modern marketplace. Consequently, some people develop strange ideas about the type of food that is good for them.

CHAP 10   *MIND OVER MATTER*

Most animals can learn to avoid poisons and to change their behaviour if they suffer a dietary deficiency. These mechanisms also exist in humans, but they are sometimes overridden. They are overridden by mental processes, based upon beliefs. Humans, unlike other animals have extraordinary mental powers that amount to 'mind over matter'. Examples such as anorexia, and hunger strike, show how powerful such mental attitudes can be. Those interested in manipulating our behaviour have not overlooked this aspect of human nature.

GLOSSARY

ENDNOTES

Preface   *The evolution of human food*

You may think that the idea of the evolution of human food is misconceived. How can food evolve when it is not alive? At least it is not alive when it becomes food. Perhaps one should say – the history of food. Not a bit of it – human food evolves because it is part of us. It is part of our ***phenotype***, the bodily expression of our genetic makeup.

In his book *The Extended Phenotype* Richard Dawkins points out that the webs that are characteristic of particular species of spider, and the 'houses' characteristic of particular species of caddis fly, are no less the products of natural selection than the animals themselves. It makes just as much sense to talk about the evolution of spiders' webs as it does to talk about the evolution of spiders.

The spider's web is part of its *extended phenotype*. It is the product of the behaviour of that particular spider and subject to same kind of genetic control as is the spider itself. In this book, I will argue that our food is part of our extended phenotype, and that it makes biological sense to talk about the evolution of human food.

Our food results from our behaviour, just as the caddis fly 'house' results from the fly's behaviour. The fly has to forage for the pebbles, and other materials, that it uses to build its house. It has to be able to recognize the materials as being suitable for building a house. Humans have to forage for their food (even if only in the supermarket), and they have to be able to recognize the materials as being suitable to eat. Other animals, also, are genetically predisposed to like certain types of food and to avoid others.

Of course, not all our food-related behaviour is under genetic control. Some of it is, and some of is the result of our culture. But it still makes sense to talk about the evolution of food. The processes involved are a mixture of genetic and cultural evolution. This is true not only for humans, but also for some other species. It may be true that the cultural element carries greater weight in humans, but this does not alter the fact that our food evolves as a result of our own behaviour. It is part of our extended phenotype.

In this book we will track the evolution of human food over the past 15000 years. During that time our bodies have not changed, but what we put into them has changed a lot. What we put into them is the result of our own behaviour, our own attitudes and our own desires. It is part of our evolution.

I thank Tom Bosser, Phil Budgell, Maeve Deign, Barbara Rolls, Colin Tudge, and Andrew White, for their helpful comments on the first draft. I also thank Penelope Farmer for putting up with food-talk.

Chapter 1   *On being hungry*

Are you hungry? In fact there are few readers of this book who will ever have experienced true physiological hunger or thirst. There are some who may have been on hunger strike, or been imprisoned with inadequate rations but most of us do not know what it is like to be really hungry or really thirsty. There are many who will never read this book who have experienced real hunger or thirst. As a result of famine, war, or residence in refugee camps, there are people whose bodies have been damaged through lack of water and/or nutrients.

You are reading this book therefore you will figure in this book. In this chapter you will experience your ***first death by eating***. In the next chapter you will be another 'historical' person, and in subsequent chapters you will inhabit a succession of bodies. Altogether you will have nine lives, and each life will end in some kind of ***death by eating***.

Fifty thousand years ago human bodies and brains were much as they are now. This is the time that Jared Diamond[1] has called the Great Leap Forward. It was a leap forward in technology. Archaeological sites dating from this time show that humans had sophisticated tools, such as needles, fish hooks, spears and harpoons, as well as works of art. This period was followed by a great geographic expansion out of Africa, into Australia, Asia, Europe and the Americas. All humans at that time were hunter-gatherers, obtaining plants and animals to eat, through their knowledge of nature and of animal behavior. Their diet was probably very varied, but they would

10

have experienced the same hunger and thirst as you and I. They were humans, but they were also animals (as we are) and they would have had much in common with other hunting and gathering animals. In this chapter we will look at those aspects of hunger and thirst that we have in common with other animals.

Your body is fundamentally the same as human bodies have been for thousands of years. What has changed over those years, and especially in the past 10,000 years is the food that you have been putting into your body. It is your food that has changed, as part of your evolution. Your food is the result of your behavior, and your circumstances. Unless you are very unlucky, your behavior determines your circumstances. Humans decide where and how to live, and over the years they have repeatedly changed their mode of living. Their habitat has evolved, together with their diet. Technically, your food is part of your *extended phenotype*[2].

Hunger for water

We can all say that at some time we have felt hungry or thirsty. So what is the difference between the hunger and thirst that we have experienced, and that of people who are starving or severely dehydrated? Let us take water first, because water is the most essential and pressing component of the human diet.

Thirst is a *motivational state* that arises primarily as a result of dehydration of the body tissues. By motivational state we mean that it is a bodily state, registered by the brain, and resulting in a *mental state* (in humans) that represents a desire to do something about the situation (e.g. find water, drink water). Animals (including humans) that are deprived of water for a long period suffer from dehydration and this leads to a type of thirst[3] that has three main effects on behavior. It increases the tendency to seek water, it diminishes the tendency to eat, and it increases the tendency to find a cool place.

11

Water is essential for the body to function, because it is the basic medium for the transport and diffusion of the chemicals that make up the body. When short of water it is important for animals to cut down on water loss from the body. Land animals lose water in the excretion of waste products, and in response to high environmental temperature. Water loss through excretion can, to some extent, be counteracted by absorption of water from the intestine into the blood stream and by cutting down on food intake. Water loss by evaporation from the body surface can be counteracted by seeking a cool place. Birds and mammals regulate their body temperature and rely on evaporation from the skin and lungs to cool themselves. Thus in response to high temperature, humans sweat and dogs and birds pant. By seeking a cool place, these water losses can be reduced.

Most land animals have water conservation mechanisms of some sort, but these can never compensate for water already lost and the thirsty animal can re-hydrate its body only by finding and drinking water. Finding water is not always easy. Humans cannot smell water, although many other animals can. However, humans can use their general knowledge to find water (knowing that water flows down hill may help) and they can also gain explicit knowledge from others (being told where to find water). Moreover, like many other animals, humans are quick to learn about cues as to the likely availability of water (there is likely to be water where there are lush green plants).

In some circumstances it is not possible to obtain water, either because there is no water in the vicinity (e.g. in the dessert) or because the person is constrained (e.g. imprisoned), or because the activity involved in getting water is considered to be too dangerous. In such circumstances the unfortunate person is likely to experience real thirst[3] that is thirst resulting from tissue dehydration. Extreme dehydration causes a drop in blood pressure, consequently less oxygen is taken in from the lungs, and to compensate for this, the breathing rate increases.

Dehydrated people experience headaches, nausea and loss of appetite. Prolonged dehydration is fatal[3].

Real thirst is an experience that is quite different from the ordinary thirst that we all have from time to time. To understand this we need to understand why we drink in an everyday situation. First of all, many animals, including humans, tend to drink when they eat. The food we eat will eventually cause water loss because the waste products have to be excreted. Eating dry food can be a cause of dehydration in the relatively short term, because water enters the gut from the blood stream to aid digestion. Moreover, much of our food contains salt, and excess salt is removed from the body in our urine. The dehydration caused by a meal taken without water is, of course, delayed somewhat. However, many people (and other animals) drink when they eat, or even just before they eat. This drinking is caused by an apparent thirst that is a purely psychological type of thirst. The water taken at meals forestalls any dehydration that may have occurred as a result of food intake. In other words, the drinking occurs in anticipation of future water loss. It is sometimes called ***anticipatory drinking***. My dog Border is interesting in this respect. (My dog is a female border terrier called Border). When she gets the idea that there is a walk or some other form of exercise in the offing, she dashes to her water bowl and has a quick drink. Similarly, if you are going to play a game of tennis, or go for a run, you may take a quick drink before you set out.

Anticipatory drinking also occurs in response to a rise in ambient temperature. In one experiment[4] pigeons were kept without water for forty-eight hours in cages kept at various temperatures, ranging from 0° to 30°C. They were then allowed to drink their fill in a test cage at 20°C. The temperatures at which the birds were kept without water made no difference to the amounts drunk in the test. In another experiment, the same pigeons were kept without water at 20° for forty-eight hours, and tested at 0, 10, 20, 30 or 40°C. The amounts drunk increased progressively with the temperature. These results show that the higher the temperature at the time of drinking, the more is drunk. Similar results have been obtained with rats[4]. The temperatures experienced during water deprivation made no difference to the amounts drunk (Note that animals tend to eat less at the high temperatures, and this aids water conservation), but the temperature at the time of drinking did make a difference.

Thus a psychological type of thirst can arise simply as a result of increases in ambient temperature. There may be no physiological dehydration, but when we feel hot, we feel thirsty and we drink in anticipation of water that we are going to lose as a result of sweating. We rarely suffer dehydration as a result of hot weather. Of course, if there is no water available, or if we are very busy, or taking a lot of exercise, then we may become dehydrated[3]. But overall, you will rarely experience real thirst. The thirst that we feel from time to time is a purely psychological thirst. When we respond to this by drinking, then we forestall tissue dehydration.

So far, we have been assuming that when we are thirsty we drink water. If the water contains other substances, such as sugar or salt, then the consequences of drinking are not so straightforward. Normally we take in sugar and salt with our food. If we also do so when we drink, then we are altering our hunger as well as our thirst. In fact, water containing sugar, salt, or other substances is not as rehydrating as pure water. So to satisfy our thirst we may have to drink more, thus altering our hunger even more. In other words, the so-called soft drinks are not soft on the body. They are hard on the body, because they usually contain unhealthy amounts of sugar (see Chapter 7). Thirst is relatively straightforward, being a *one-dimensional motivation*. You can want more or less water, but you cannot want different water. However, things become complicated when you take in other substances with your water.

Hunger and energy

Animals require food that can be digested to provide energy and certain specific nutrients and vitamins. The food is digested and the proceeds of digestion enter the blood stream. In the process of *metabolism* energy is made available to the cells of the body primarily in the form of glucose dissolved in the fluid surrounding the cells (*extracellular fluid*). In this process, water, carbon dioxide and heat are produced as byproducts. The glucose in the extracellular fluid may come directly from digestion or may be released by the liver from its store of glycogen. In other words, the products of digestion are either utilized right away (glucose) or stored as glycogen (in the liver) or fat.

In the absence of digestive products, the cells of the body can obtain energy in the form of glucose (from the liver) or fatty acids (from stored fat). The cells of the nervous system are the exception for they can utilize only glucose. For this reason it is important that the glucose level in the blood be maintained within fairly narrow limits, and that the availability of glucose is controlled by the hormone *insulin*[5]. The liver and the pancreas are the key players here. Together they regulate the blood glucose levels.

A feeling of hunger is experienced when the glycogen level of the liver falls below a threshold. This information is passed to part of the brain called the *hypothalamus*, via receptors in the liver. The sensation of hunger typically begins after a couple of hours without eating, but this is not real hunger because the situation is usually rectified by the release of stored glucose (i.e. from glycogen). If you are expecting a meal at a particular time but have to miss it for some reason then you are

15

likely to feel hungry, due to a drop in blood glucose levels. This happens because when the liver is 'expecting' a meal (i.e. expecting the products of digestion to enter the blood stream). The liver reduces the rate of conversion of glycogen into glucose. The liver is very susceptible to **conditioning** by external stimuli. If you ring a bell every time a dog (or other animal) is fed, then the liver 'learns' to reduce its glucose output at the sound of the bell. The same applies to time-of-day. The liver learns to reduce glucose output at particular times of day. This phenomenon is particularly marked in people who have their meals at regular times. If the liver did not react in this way, then the blood glucose levels would rocket soon after a meal, because sugars are absorbed quickly into the bloodstream. Here we have another example of anticipatory control.

Some people experience hunger pangs. These are due to muscular contractions in the stomach. Hunger pangs usually do not begin until twelve to twenty-four hours after the last ingestion of food. A single hunger contraction lasts about 30 seconds, and pangs continue for around thirsty to forty-five minutes, then hunger subsides for around half an hour to one and a half hours. These pangs are signals indicating that the stomach is empty. They are not normally indicators of starvation, because the body may have plenty of reserves. Some scientists consider that starvation does not really start until all carbohydrate and fat reserves are utilized, and protein breakdown begins. Hunger for energy (glucose) is similar to thirst, in being a **one-dimensional motivation**. However, in addition to hunger for energy, animals (including humans) may have cravings. They may have a hunger for specific components of the diet.

Hunger for Specifics

Humans are omnivores, and eat a wide variety of types of food. Whereas water is a simple homogeneous substance, H_2O, food is very complex and variable. Whereas our brain can measure the state of dehydration of the body, it cannot possibly measure, directly, all the complex nutrients and elements involved in hunger. Nevertheless we do feel hungry from time to time, so our brain must be telling us something. Let us look first at an animal that is not an omnivore.

The bear-like Koala is a marsupial mammal that lives in Australia, and feeds exclusively upon Eucalyptus leaves. Its food is pretty much homogeneous. It can eat more or eat less, but it cannot vary the diet. This means that the animal's hunger exists in one dimension, like thirst. The problem with a homogeneous diet is that you have to take the bad with the good. At certain times of the year the young leaves and shoots of the Eucalyptus contain hydrocyanic acid, which is toxic. In the wild, Koalas switch to trees that do not have fresh sprouts but in captivity they are sometimes not given this choice, and they may not survive. Obviously, if an animal is to thrive on a very restrictive diet, it must have special physiological adaptations to enable it to cope with that diet.

Most animals have a more varied diet and are able to adjust their intake to the needs of the body. In other words, in most animals hunger is a *multi-dimensional motivational state*. When we are really hungry, we may be lacking energy, but we may also be lacking specific nutrients. If we are starving we may be willing to eat anything taking the good with the bad. On the other hand, people are often malnourished without being starved, and this may result in cravings, sickness, etc.

Suppose that you are well nourished, apart from the fact that your body has a salt deficiency. Common salt, $NaCl$, is involved in many bodily processes, and lack of it can have serious consequences. Symptoms of salt deficiency in humans include nausea, vomiting and headaches. If the situation worsens then convulsions and coma may occur[6]. Mild salt deficiency often leads to a craving for something that tastes salty.

Most animals are able to taste salt. There are receptors on the tongue that detect salt. So a normal person can detect the presence of salt in food, provided the concentration is above a certain threshold. The flavor of food depends upon both taste and smell. This can be demonstrated by asking a person to close his or her eyes and tell the difference between small pieces of apple or onion placed on the tongue. Most people find the discrimination easy if they are allowed to breathe through the nose, but impossible if they are asked to hold their nose while taking the test. The point here is that the salt deficient person will crave a salty taste and will associate this with certain food items, such as potato crisps, that they have encountered in the past. As we shall see, the salt taste is readily subject to *conditioning.* This means that people associate the salty taste with enjoyable and rewarding food and they become used to it. People who do not have the salt habit have less of a salt craving, are much better at detecting the presence of salt in their food than those who usually add salt. In other words, food that tastes salty to the people who are adjusted too little salt in their diet, does not taste salty to people who are used to a lot of salt in their diet.

In northern medieval towns there were long periods when no fresh meat or fish was available. It was difficult to keep animals alive and healthy throughout the northern winter, and animals were slaughtered in the autumn and the meat preserved by salting or drying. Fish could be transported inland only if it was similarly preserved. The result was that that the preserved food had to be laboriously desalted before it could be eaten. Inevitably much salt remained, and we can tell from recipes of the time that great efforts were made to disguise the saltiness of meat by means of anti-salt sauces made from salt-absorbing dried legumes or frumenty, plus whatever spices were available[7]. The result was that many people were forced to eat more salt than was good for them[8]. In other words, to obtain sufficient protein, medieval townspeople had to eat the bad with the good.

Lack of salt in the diet is bad for health[6], and so is lack of thiamine (vitamin B_1)[9], but there is an important difference. The difference is that people can taste salt, but they cannot taste thiamine. Rats can also taste salt, and cannot taste thiamine. Nevertheless, rats can discover foods that contain thiamine. How is it done?

The psychologist Paul Rozin has investigated specific hungers in rats. Most animals are unable to detect many essential vitamins and minerals either by taste or by their levels in the blood. Nevertheless, animals deficient in certain vitamins or minerals do develop strong preferences for foods containing the missing substances. For many years such specific hungers posed something of a problem for scientists trying to explain how animals knew which food contained the beneficial ingredients.

Paul Rozin and his co-workers[10] showed that rats deficient in thiamine develop an immediate marked preference for a novel food, even when the new food is thiamine deficient and when the old food has a thiamine supplement (which of course they cannot detect). If the deficiency persists, the preference is short-lived. On the other hand, if consumption of novel food is followed by recovery from dietary deficiency, then the rat rapidly learns to prefer the novel food. Such rapid learning on the basis of the physiological consequences of ingestion enables the rat to exploit new sources of food and thus find out which contains the required ingredients. Rozin realized that a vitamin deficient diet is like a slow acting poison. He noted that an aversion to a familiar but deficient food persists even after the rat had recovered from the deficiency. Rats (and other animals) become sick after eating a poisoned food, and also

19

show an aversion to such food and show a more than normal interest in novel food.

In other words, the thiamine deficient rat feels ill. If it eats a novel food and then starts to feel better, it associates its recovery with the taste and smell of the novel food, and not to other things that it experienced while ill. In some of Rozin's experiments the novel food contained no thiamine (the thiamine was injected directly into the animal's body), but the rats nevertheless associated the novel food with recovery. So the rat suffering from a dietary deficiency tends to sample novel food more than usual, and rapidly learns which food is good. Humans have similar mechanisms. In the days when many people had coal fires, doctors were often consulted by parents whose young children seemed to be sucking on coal. It turned out that such children were suffering from lack of certain minerals, and had discovered by trial and error that coal made them feel better. Toddlers habitually put things in their mouths, and coal contains quantities of selenium, copper, and iron: all minerals that are necessary (in small quantities) for a healthy diet[11].

There are some twenty vitamins and minerals that are known to be essential for good health[12]. These are the *essential nutrients*. A diet that is deficient in one or more of these leads to what is called a *specific hunger*. As long ago as the 1940's the psycho-physiologist Curt Richter indicated that there is a kind of harmony amongst the animals' physiological and behavioral responses. Rats left to their own devices with a large variety of purified nutrients from which to choose select a balanced diet from this cafeteria. Of course, the animal faced with food that is deficient in some respect has problems, and similar problems arise with food that is contaminated in some way. We have already seen how medieval townspeople, in the northern winter were obliged to eat meat that contained excessive amounts of salt. Such people required protein in their diet, and to obtain sufficient protein they needed to eat salted meat or fish. Excessive salt is bad for health, but protein deficiency is also bad. So what to do?

Somehow there has to be a compromise. Suppose that food X contains protein and salt in fixed proportions . If this is the only food available, then the consequences of eating are one-dimensional (i.e. you can eat more or less food, but you cannot eat different food). In other words, to obtain a certain amount of units of protein it is necessary to take in an excessive amount of salt. A possible alternative would be to eat the amount of food that provides the right amount of salt, but this would provide too little protein. Something between these two extremes would seem to be called for - a compromise or *tradeoff*. In other words an excess of protein is traded for a deficiency of salt or vice-versa[13].

Dietary deficiencies occur in humans and other animals when they are obliged to live in an environment that is less than ideal nutritionally speaking. A good example is lack of vitamin C. Insufficient intake of vitamin C leads to scurvy, an unpleasant and ultimately fatal disease. Scurvy leads to the formation of liver spots on the skin, spongy gums, and bleeding from the eyes, nose, etc. Scurvy was at one time common among sailors on ships that were out to sea for longer than perishable fruits and vegetables could be stored. Fruits and vegetables contain quantities of vitamin C, and it was eventually realized that scurvy could be prevented by making sure that the sailors' diet included vitamin C. The credit usually goes to Captain James Cook, but he was preceded by James Lancaster who was in charge of the first expedition of the embryo East India Company in 1601. He provided the sailors on his ship with a daily dose of lemon juice, and he noted that his men were free of scurvy, while those on the other ships in the expedition suffered badly[14]. In 1614 John Woodall , Surgeon General of the East India Company, published his book *The Surgeon's*

Mate, a handbook for apprentice surgeons aboard the company's ships. In it he described scurvy as resulting from a dietary deficiency. His recommendation for its cure was fresh food or, if not available, oranges, lemons, limes and tamarinds. By insisting on an appropriate diet, James Cook succeeded in circumnavigating the world (1768-71) without losing a single man from scurvy.

Recently, scientists[15] have discovered that captive animals, on farms or in zoos, may suffer from dietary deficiencies. Captive adult ungulates (horses, cattle, and giraffes) often show stereotypic oral behavior, such as chewing inedible objects. This is more common if the animal is fed a concentrated low-fiber diet, or a diet deficient in certain minerals. In other words, unnatural foraging regimes lead to unnatural behavior. The captive animal does not have the *opportunity* to forage in a natural way. In nature it would be able to pick and choose amongst a variety of types of food. So if your horse chews the stable door, you might consider looking into its diet.

Hunger and opportunity

You may feel hungry, and food is readily available, but you are busy at work, or you are engrossed in a book. Just because you are hungry and food is available, does not mean that you must eat now. You may prefer to do something else. It is all a matter of *motivational priorities.*

Let us take a particular example: A herring gull is sitting on its nest incubating its three eggs. Normally the parents take it in turn to incubate because a nest left unattended is likely to be raided by predators. Usually the partner leaves the nest to forage for food and returns within a few hours. Sometimes however the return may be delayed as a result of mishap or injury or capture by a scientist. What should the sitting bird do? On the one hand its partner may return at any time, but on the other hand the sitter becomes increasingly hungry as time passes. Eventually the sitting bird should quit the nest to search for food. Because herring gulls breed for many successive seasons, it is not in the sitter's genetic interest to endanger its own life for a single clutch of eggs. The question is at what point should the sitting bird decide to quit?

If the incubating bird quits too early, it may endanger its clutch unnecessarily because its mate may have merely been delayed. If it quits too late, it may endanger its life unnecessarily because it may not be able to find food as quickly as usual, or it may be so hungry that it takes unnecessary risks while foraging. To determine exactly when the bird should quit, these alternative risks have to be evaluated quantitatively, an evaluation carried out effectively by nature in designing the animal to make the most appropriate decisions.

To investigate this type of problem, it is necessary to make field measurements of the risks run by the animals concerned. In this case we would need to discover the risks to the eggs of being left unattended and the risk to the incubating bird of being left without food. Biologist Rudi Drent found that, in the Friesian Islands 20% of the herring gull eggs were lost to predators, and 12% succumbed to the physical hazards of incubation. On Walney Island, Cumbria, biologists Robin McCleery and Richard Sibly placed weighing machines under the herring gull nests. They then caught and marked the parent birds and measured their skeletal parameters. The amount of fat a bird is carrying can be estimated from its weight in relation to its skeletal size. The amount of food an animal obtains from a foraging trip can be estimated from the change in weight measured on the nest before and after foraging. The quality of the food can be estimated from the faeces gathered from around the nest. They discovered that the feeding preferences of individual birds were very important. From their experiments and calculations they concluded that the members of a mated pair should have complementary food preferences, that energy reserves should be maintained between 500 and

1200 kilocalories, and that the parent should desert the nest if the energy reserves fall below 200 kilocalories[16].

Now let us look at it from the point of view of the individual bird. The sitting herring gull is hungry, and is motivated to forage for food. On the other hand, it is also motivated to sit on the eggs to incubate them and to protect them. The bird cannot both incubate and forage so (as hunger increases) is has to come to a ***decision***. During this process there is a balance of motivational priorities, a continual ***trade-off*** between sitting and foraging. Scientists are able to (mathematically) work out such tradeoffs, and come up with quantitative explanations of how particular tradeoffs work. In the present case, the decision to quit the nest depends upon the amount of fat the bird is carrying. So the decision depends upon at least two internal variables – the animal's hunger and its fat reserves. In addition to information about its hunger state, the animal will obtain information about the likelihood of obtaining food if it were to go foraging. It will assess the feeding opportunities. Such factors as the time of day, the state of the tide, etc. will influence its decision. There would be no point in quitting the nest in the middle of the night when the chances of obtaining food were negligible. As we shall discover in Chapter 3, the foraging animal leads a complex life.

How do animals assess foraging opportunities? Much depends upon the style of foraging. For some sit-and-wait predators the opportunities are fairly obvious – either the prey comes within reach, or it does not. The stimulus indicating possible prey is often an all-or-none matter. For example, physiologist Jerry Letvin showed that photoreceptors in the frogs' eye (retina) are connected together in such a way as to form 'bug detectors'. These respond preferentially to small dark moving objects against a background of non-moving objects[17]. In nature such dark moving objects are usually flies. The frog detects the fly and quickly grabs it with its tongue.

Predators that actively search for food take into account numerous aspects of the external environment, as we shall see in Chapter 3. These include cues to the likelihood of obtaining food, the apparent palatability of the food, etc. All these external factors are classed by psychologists as aspects of *incentive motivation*, and by biologists as *cue strength*. The point here is that the psychological aspect of hunger, (the motivation to seek food) depends upon both the internal hunger and the external food stimuli. So if I am moderately hungry and I am walking along the street to buy a book I may pass a shop with an appetizing display of food, cakes, pies, etc. The sight of this food immediately increases my (psychological) hunger. My motivation to eat is increased. It may even increase so much that I decide to buy some food instead of buying a book.

So being hungry is not simply a matter of my internal state. In fact I am probably rarely really hungry in this respect. It is a matter of the combination of my internal state, and other factors such as the time of day, the sight of food, and so on. Whether I do something about my hunger depends largely upon other aspects of my motivational state, my desire to buy a book, my willingness to make the effort (i.e. pay for the food in some way), and my time budget. Do I have the time to wait in the long queue to buy the attractive food displayed in the shop window? In my case, the answer is probably no. I am impatient, and time is important. As we shall see in Chapters 3 and 6, in deciding when and what to eat, time is a major player.

Now what about you? How have you fared? You are a sixteenth century mariner. You are, in effect a captive animal. You are on board *Penelope*, a sailing vessel that, together with two others under the command of James Lancaster is hoping to join the spice trade out of the East Indies. The three vessels sail from Plymouth in the spring of 1591. They soon reach the Canaries and then set off for Cape Verde before crossing the equator. They followed the trade winds towards South America before turning east towards the Cape of Good Hope. Then they hit the doldrums and their provisions are running low. They are reduced to biscuits and salt pork, a monotonous diet. They have now been at sea for three months without eating any fresh fruit or vegetables. The men begin to sicken. Their teeth drop out and purple splodges appear on their bodies. Their joints stiffen and thin streams of blood come from their eyes and noses. Many of them die, including you. They die not from starvation but from scurvy, the result of a specific hunger that they are not able to remedy. It was not your fault because you were a captive animal and you had no opportunity to vary your diet in a way that would have been natural for you as an omnivore. It is our first case of *death by eating*, but you would not have survived anyway. After rounding the Cape of Good Hope a tremendous storm sank the *Penelope* with the loss of all hands[18].

Chapter 2

The evolution of feeding strategies

In the natural world, to earn a living as an animal, you have to do your own thing, and do it well. What you do for a living depends partly upon where you live. Some species, such as humans and rats, have colonized large parts of the globe, but most species live in a more restricted **habitat.** Each species has a specific **tolerance** range of environmental factors, such as temperature, humidity, etc. This depends largely upon the animal's physiological mechanisms, and its ability to adjust to environmental changes by means of appropriate behavior. So an animal's habitat is usually described in terms of the salient physical and chemical features of the environment, and these are largely determined by climate.

Where I live in the Canary Islands is classified as a desert habitat, because of the high average temperature and low rainfall. The local lizards make a living by eating plants and insects, and avoiding predators. When the weather is hot they can move very fast, and my dog Border is unable to catch them as they dart for cover. When the weather is cooler, she can catch them if she can find them. The lizards are well camouflaged and hide when the ground temperature is low. Because they are not warm-blooded, the lizards do not use much energy when they are stationary, and this means that they can survive on much less food than a warm blooded bird or mammal of comparable size.

The association of animals and plants that live in a particular habitat is called a *community*. The members of a community are usually divided into producers, consumers and decomposers. The plants are producers, obtaining their energy from sunlight, the lizards are primary consumers, because they eat plants. They are low down on the *food chain*, being preyed upon by kestrels and shrikes. Kestrels eat what they can get, on the spot, or at home with their young. Shrikes sometimes store food, impaling lizards on cactus spines.

The lizards occupy a particular niche. They eat green plants and insects, and they themselves are eaten by birds. A *niche* is the role played by a species in the community, in terms of its relationship both to other organisms and to the physical environment. Thus an herbivore eats plant material and is usually preyed upon by carnivores. The species occupying a given niche varies from one part of the world to another. For example, a small herbivore niche is occupied by rabbits and hares in northern temperate regions, by the agouti and viscacha in South America, by the hyrax and mouse deer in Africa, and by wallabies in Australia.

When animals of different species use the same resources or have certain preference or tolerance ranges in common, *niche overlap* occurs. Niche overlap leads to competition for resources. Two species with identical niches cannot live together in the same place at the same time when resources are limited. This is called the *competitive exclusion principle*. So where niche overlap occurs, the two species must play slightly different roles in the community. On my land, the prime herbivores are rabbits. To some extent the rabbits and lizards are in competition, especially during the dry season when there are few green plants. The main differences between the two species are that the rabbits forage at night and the lizards forage by day. The rabbits can dig up the roots of plants and the lizards cannot do this. The lizards eat insects whereas the rabbits do not.

Lizards have been around for a very long time – much longer than humans. So they must be doing something right. They must have a good *evolutionary strategy*. Such strategies are deployed by plants and animals, but they are not thought up by individuals. An evolutionary strategy is a passive result of *natural selection* that gives the appearance of a ploy employed by genes to increase their numbers at the expense of other genes. To understand this we need to look at *evolutionary theory*.

Evolutionary strategies

Evolution is a process by means of which species characteristics originate by development from earlier forms. It is now widely acknowledged to be due to *natural selection*. The theory that evolution results from the process of natural selection is due to Charles Darwin (1809-1882). The elements of Darwin's theory can be stated as follows: Within any population of organisms of the same species, there is considerable variation among individuals. Much of this variation is inherited. Many more individuals are born (in the case of animals) in each generation than survive to maturity. Therefore the likelihood of an individual surviving to maturity will be affected by its particular traits, especially those that it has inherited from its parents. If the individual survives to maturity and reproduces successfully, its offspring will tend to perpetuate the inherited traits within the population.

Evidence for evolution comes from the fossil record, comparison of present-day species, the geographical distribution of species, and some observations of evolution in action. Animal behavior can only be fully understood in terms of its evolutionary history, and in terms of the role it plays in the survival and reproduction of the animal. However, the evolution of animal behavior is difficult to study directly, because there is no fossil record, but much can be learned by comparison of species. Scientists interested in the evolution of behavior have developed a body of *evolutionary theory* that enables them to develop hypotheses concerning the evolution of behavioral traits.

Evolutionary biologists are interested in explaining how a state of affairs observed today (such as the typical behavior of a certain species) is likely to have come about as a result of evolution by natural selection. To account for the establishment of a particular genetic trait, they imagine a time before the trait existed. Then they postulate that a rare gene arises in an individual by mutation, or arrives with an immigrant, and that individuals carrying the gene exhibit the trait. They then ask what circumstances will favor the spread the gene through the population. If the gene is favored by natural selection, then the individuals with genotypes incorporating the gene will have increased fitness. The gene may be said to have invaded the population. To become established a gene must not only compete within the gene pool, but must also resist invasion by other mutant genes. It is as if the genes develop a strategy to increase their numbers at the expense of other genes. Thus an *evolutionary strategy* is a passive result of natural selection that gives the appearance of a ploy employed by genes to increase their numbers at the expense of other genes.

An evolutionary strategy is a strategy deployed by the individual. For example, a gene that makes an insect unpalatable to predators may not spread in the population, because by the time the predator discovers that the prey is distasteful, the insect is dead. However, if the gene also provides the insect with distinctive *warning* coloration, then the predator will more easily learn to avoid other insects carrying the gene, and the gene would increase in frequency in the population. In other words, if you are an animal that is poisonous, or has a nasty sting or bite, then it is a good evolutionary strategy to advertise the fact. After a bad experience, a predator soon learns to avoid a distinctive-looking animal. Now, the warning coloration is part of the animal's *phenotype*. It is an aspect of its morphology and behavior. What we are interested in, and the reason we are delving into the idea behind evolutionary strategy, is the human phenotype, particularly as it relates to our food. However, that is for later. What we need to do now is to understand some of the complexities of evolutionary strategy.

It is sometimes of advantage for different species to share the same warning signals. For example, some social wasps, and some solitary wasps and the cinnabar moth caterpillar, are bad news to predatory birds, and they all have black and yellow warning coloration. Birds that have learned to avoid cinnabar caterpillar will not attack wasps, but birds with no experience of wasps or cinnabar caterpillars will attack wasps. The predators need to learn only one pattern in order to avoid all the species in the assemblage. This is an example of a good evolutionary strategy, because a gene favoring black and yellow coloration benefits the bearer simply because this *warning* signal is commonplace, and therefore a predator is likely to have already learned about it. Similarly, we quickly learn to identify police cars, whatever country we are in.

In general, the survival of a trait within a population depends upon the extent to which the trait contributes to *reproductive success*, which depends partly upon the selective pressures inherent in the environment (food availability, predators, etc). This is as true for plants as it is for animals. Plants gain their energy by photosynthesis, a process that requires sunlight and a suitable photochemical, usually chlorophyll in green plants. Animals that eat the green parts of a plant are depriving that plant of its energy source, and plants have evolved various strategies of deal with this problem. Some store energy underground, examples being potatoes, turnips, etc. If their leaves are eaten they can regenerate them. Others have spines and thorns that deter some predators, examples being cacti and thorn-bushes. Others manufacture chemicals that are poisonous to animals. However, not all plants are anti-animal. Many exploit animals as an aid to reproduction.

Thus many flowering plants manufacture food for insects and advertise their wares. The insects are attracted by the bright flowers and home in on the nectar. Matters are so arranged that the insect unwittingly transport pollen from one flower to another, thus facilitating cross-fertilization of that particular species of plant. It is of advantage to plant genes that the seeds be dispersed widely, and many plants recruit animals to help in this process. Some manufacture fruits that animals like to eat, and others have fruits that cling to an animal's fur. In many cases there has been *co-adaptation* involving particular species of plant and animal, to the benefit of both.

When we come to animals, there are a number of features of the environment that could jeopardize reproductive success by leading to the death of the parent by starvation, or predation; or failure to breed as a result of competition for mates or nesting sites; or failure of the young to survive due to lack of parental care, food, or protection from predators. Animals have evolved a variety of strategies to mitigate these problems. The evolutionary strategies that concern us here are those that relate to their food. So let us start with their plant food.

Living with plants

Ultimately, all the energy that we put into our mouths comes from plants. Plants obtain their energy primarily from sunlight, and when an animal eats plant material it obtains its energy from the plant. From the (evolutionary) plant point of view, there are advantages and disadvantages of being eaten by animals. The leaves, flowers, fruits and roots are all targets for

32

some animal or another, and the strategies employed by plants are related to these targets.

Let us look first at grazing animals[1]. Grass is eaten by mammals, such as rabbits, rodents, kangaroos, and ungulates (the large hoofed mammals); and by birds, such as geese; reptiles, such as tortoises; and some invertebrates. The latter include snails and slugs, grasshoppers and caterpillars, and some termites. Natural grasslands are made up of perennial grasses that propagate by lateral stems spreading over and under the soil surface, in an extensive ramifying system. The growing points (intercalary meristems) are on or under the soil surface, where they are protected from trampling, grazing and fire. Grazing stimulates the development of these into new shoots, but prevents the production of flowering stems. Consequently the nutrient content of grassland is improved by grazing. The soil receives nutrients in the form of animal faeces, and the mechanical effects of grazing promote new leaf growth.

The new leaves are especially palatable to animals, having less cellulose than old leaves, and more protein and soluble carbohydrate. So anything that promotes new growth increases the nutrient content of the grassland. Although the net productivity of grassland is less than that of forest, more than half the annual production is available to grazers, whereas only ten percent of forest production is available to such primary consumers.

The evolutionary strategy followed by grasses involves encouraging grazing. In this way the grass is kept young and vigorous. Grasslands that are not grazed, age quickly and produce flowering stems. The flowers then produced seeds that are dispersed by wind. In this way a particular species of grass has a chance of colonizing new territory. But grasslands that are not grazed are invaded by other plants that employ different evolutionary strategies. This results in an *ecological succession*, in which one dominant species is replaced by another. Typically, under-grazed grassland reverts to scrubland and eventually to woodland.

Grasslands are maintained by grazing and grazers were important throughout their evolution. The first ungulates are thought to have been forest animals, feeding mainly by browsing. By eating seedling trees they opened up woodland, creating clearings where grass could grow. Fossil evidence indicates that this trend started about 25 million years ago, and during that period the grazing animals exerted selection pressure favouring plants that could survive despite grazing. When humans came into the picture they accentuated the trend by developing a pastoralist lifestyle, keeping herds of semi-domestic grazing animals. In addition, humans felled forests and caused fires thus increasing the grassland at the expense of forest. Thus grasslands have co-evolved with their animal inhabitants. Similar evolutionary stories can be told about browsing animals.

Many species of animal are browsers, eating the leaves of trees and bushes. They include mammals, such as giraffes, elephants, and sea cows (manatees); birds, such as grouse and pigeons; lizards; and numerous species of insect, including beetles, grasshoppers and above all, caterpillars. Trees have evolved many kinds of defence against such predators. As we shall see, some predators are deterred by spines and thorns, others by toxic chemicals.

Trees also exploit animals as aids to cross-fertilization and seed dispersal. Trees are flowering plants (apart from the conifers), and reproduce sexually. Self-pollination is a form of in-breeding that can lead to genetic degradation, so most trees are out-breeders. To achieve this they must evolve a way of getting their pollen (male gametes) to the female reproductive organ (style/stigma and ovary), preferably on another tree. In general trees reject the pollen of other species, and some, such as plums and apples, also reject pollen that is too genetically similar to themselves. Some trees rely on a particular species of animal to do this for them. For example, each species if fig, and there are about 700 of them, is pollinated by just one species of fig wasp. The fruit of a fig tree is a case (called the syconium) that encloses male flowers and the female flowers. So the syconium entirely encloses the reproductive organs. In other words, the flowers are hidden from the outside world.

Figs are important plants in tropical forests, and they also occurs in subtropical and temperate areas. The relationship between fig and wasp has a very long evolutionary history. Pollen dispersal in figs is completely dependent on fig wasps, which enter the syconium where viable seeds are produced. Fig pollinators are uniquely dependent on the fig for the completion of their life cycle. Each fig species attracts its own species of fig wasp by emitting specific volatile chemicals (*pheromones*) during their receptive period and these wasps lay their eggs in the developing ovary of the host fig species. The female wasps simultaneously carry pollen from other syconia and pollinate figs actively or passively, and they show peculiar morphological and behavioral adaptations to figs and a life cycle highly synchronous with the fig flowering cycle. In other words the figs and the wasps have co-evolved a mutual relationship. They provide a good example of **symbiosis**[2].

Many flowering plants produce nectar, at some cost to themselves. Nectar contains the sugars, glucose, fructose and sucrose, and also some amino acids. So it is a highly beneficial food for animals. The provision of nectar is an evolutionary strategy to encourage visits by pollinators. For the strategy to be successful, the plant must advertise its wares. This is the primary function of conspicuous flowers. Advertisements are a form of *communication* designed (by natural selection) to induce animals to spend time and energy on behavior that they might not do otherwise (see Chapter 6). Such communication may be broadcast or targeted. Broadcast communication advertises the fact that the flower is ready for visits by pollinators. It may take the form of an attractive perfume or *pheromone* or of conspicuously colored petals. Targeted advertisement is designed to attract particular customers. Thus flowers with readily accessible nectar are often white or yellow, and are visited by short-tongued insects such as flies and beetles. Flowers with concealed nectar are usually red, violet or blue, and are visited by long-tongued insects, such as hoverflies, bees and butterflies. By providing readily accessible nectar the plant is appealing to a wide variety of pollinators, but may also be visited by destructive non-pollinators such as birds (e.g. bullfinch). By attracting particular sorts of insects to a relatively inaccessible source of nectar, the plant is able to provide floral mechanisms that are intricate and precise, thus increasing the reliability of pollination. For example honeysuckle is visited by night-flying hawkmoths with very long tongues. The flower is tubular, pale in color, and opens at dusk. It is thus unlikely to be visited by daytime marauders.

Both the broadcast and targeted pollination strategies have their benefits and risks. The benefit of broadcasting is that is has a wide appeal, although risking clumsy or destructive visitors. The benefit of the targeted strategy is increased reliability per visit, at the risk of few visits. Thus if the hawkmoth population were to decline, then the honeysuckle would suffer. As we shall discover in Chapter 6, the same principles apply to human advertising.

Insects are not the only flower pollinators. Field mice, bats and birds can also act as pollinators. Bats that lap nectar have a long snout and tongue. They become dusted with pollen as they feed. Hummingbirds, honeycreepers, and some other birds visit flowers to obtain nectar. Bird-pollinated flowers are often brightly colored, and have no scent. Pollination leads to fertilization and the fertilized ovary develops into a fruit. It is often in the interests of the plant (i.e. it is an evolutionary strategy) to make the fruit attractive to animals.

Animals play a very important role in dispersing seeds, especially in tropical rain forests. Some seeds cling to the fur of mammals, such as monkeys. Other are swallowed when animals feed on fruit, and are later dispersed unharmed in the droppings of fish, reptiles, birds, bats and other mammals. For example, the fruits of the nutmeg tree are eaten by the magnificent bird of paradise. The outer fruit is digested but the inner seed (the nutmeg) is protected by a thin hard shell, and is passes through the bird and is dispersed throughout the forest. Many edible fruit advertise their presence by pheromones. Tigers have a passion for the foul smelling fruit of the Durian tree. They disperse the seeds in their droppings, as do monkeys, pigs, tapirs and rhinoceros. It is in the interest (i.e. an evolutionary strategy) of such trees to attract animals that will eat their fruit and disperse the seeds. The animals obtain food, so the arrangement is of mutual benefit. However, not all plant-animal relationships are mutually beneficial[3].

Coping with the enemy

Many plants defend themselves from predation by producing toughened leaves, spines, and toxins that are either poisonous or distasteful to animals. Some of the poisons are deadly. Foxgloves produce several deadly chemicals, namely cardiac and steroidal glycosides. Ingestion can cause nausea, vomiting, hallucinations, convulsions, or death.

Many animals that feed upon plants have evolved strategies for countering plant anti-predator devices. Some herbivores have evolved biochemical counter measures to plant poisons, such as enzymes that counter and reduce the effectiveness of the numerous toxic substances produced by plants. A fairly common evolutionary strategy is to appropriate and make use of plant toxins. Some butterflies lay their eggs on passion flower vines. Their caterpillars feed upon the leaves that contain toxins. The adult butterflies incorporate these chemicals, making themselves poisonous to birds. The butterflies are distinctively colored and the birds soon learn to recognize and avoid them. So these butterflies advertise the fact that that are poisonous (an example of warning coloration). Other non toxic butterflies, have evolved to *mimic* this color pattern, and so benefit from the warning coloration.

The majority of leaf-eating insects are palatable and relies on camouflage for protection, so to evolve bright coloration would seem to be a dangerous strategy. However, clearly the benefits outweigh the danger, and warning coloration has evolved more than once. Thus the caterpillars of monarch butterflies feed on milkweed plants containing poisonous chemicals. They store the toxins without harm and develop into brightly colored butterflies. These also have their mimics that are perfectly palatable, but pretend to be poisonous. Of course, this ploy will work only if the mimics are not too numerous. A predator that eats a brightly colored palatable insect will not quickly learn to avoid such prey, and the advertisement will misfire.

Herbivores are unable to digest complex cellulose and rely on internal symbiotic bacteria, fungi, or protozoa to break down cellulose so it can metabolized. Symbiotic microbes also allow herbivores to eat plants that would otherwise be inedible. They detoxify the plant metabolites. For example, fungi living on cigarette beetles produce detoxification enzymes to get rid plant toxins[4]. Symbiotic microbes also assist in the acquisition of plant material by weakening a host plant's defenses. As an

example, several species of bark beetle introduce blue stain fungi into trees before feeding.[4] The blue stain fungi cause lesions that reduce the trees' defensive mechanisms and allow the bark beetles to feed.

Many herbivores have evolved behavioral strategies to counteract plant poisons. One such strategy has to do with timing. The swallowtail butterfly lays its eggs just as the young leaves of its citrus host begin to appear. The caterpillars feed on the new leaves before they develop a protective waxy cuticle. Another strategy is to immobilize plant defenses. Caterpillars of that feed on the spiky plants may spin a silken web over the spines. They crawl over this to feed upon the unprotected leaf edges.

Herbivores have evolved a diverse range of physical structures to facilitate the consumption of plant material. To break up intact plant tissues, mammals have developed teeth that reflect their feeding preferences. For instance, mammals that feed primarily on fruit, or on soft foliage, have low-crowned teeth specialized for grinding foliage and seeds. Grazing animals that tend to eat hard, silica-rich grasses have high-crowned teeth, which are capable of grinding tough plant tissues and do not wear down as quickly as low-crowned teeth. Insect herbivores have evolved a wide range of tools to facilitate feeding. Often these reflect the feeding strategy and preferred food type. For example, moths that eat relatively soft leaves are equipped with incisors for tearing and chewing, while the species that feed on mature leaves and grasses cut them with toothless snipping mandibles.

How do humans fit in?

Fifty thousand years ago there was more than one humanoid species. In Europe there was *Homo neanderthalensis* and there was *Homo sapiens*. The former was largely a carnivorous predator, and the latter an omnivore. They certainly overlapped, but it is uncertain whether they interbred (Note that if it is discovered that they did interbreed and had viable offspring, they might not continue to be classified as different species). Neanderthals had many adaptations to a cold climate, such as large braincase, a robust build, and rather large noses (to warm inhaled air). Their cranial capacity was larger than modern humans, and their skeletons indicate that they were considerably stronger. Some Neanderthals had red hair and pale skin color, indicating that they were adapted to northern (less sunlit) conditions.

Neanderthals survived in Europe until about 40-30,000 years ago. They performed many sophisticated tasks which are normally associated only with our species. For example, it is known that they controlled fire, constructed complex shelters, and skinned animals. They had spears, made of long wooden shafts with stone spearheads, thought to have been thrusting spears, and they had stone knives and scrapers[5]. Neanderthals overlapped with humans for about 15000 years. They were probably did compete with *Homo sapiens* for food. Although Neanderthals hunted large animals, such as rhinos and mammoths, whereas *Homo sapiens* probably depended more on small animals and plants. Prior to the Neolithic revolution and the invention of farming (see Chapter 4), humans (*Homo sapiens*) were hunter-gatherers. Their evolutionary adaptations included intestinal adaptations to an omnivorous diet, and the ability use tools and fire.

Humans are omnivores. This does not mean that they can eat anything, but they can eat many types of food. Humans cannot eat grass, because they cannot digest cellulose. They lack the special alimentary adaptations enjoyed by most grazing animals. The human invention of cooking may have been particularly helpful in overcoming many of the defensive chemicals of plants. (see Chapter 4) . Like other animal species, humans must have been expert foragers, and in the next chapter we will explore what this entails. It is also possible that humans entered into symbiotic relationships with other species. Even today, humans are recruited by other animals to help in obtaining food. The honey badger lives in symbiosis with a small bird called the honey guide. When the bird discovers a hive of wild bees, it searches for a badger and leads it to the hive by means of a special display. Protected by its thick skin, the badger opens the hive with its large claws and feeds on the honeycombs. The bird feeds upon the wax and bee larvae, to which it could not get access unaided. If the honey guide cannot find a badger, it may attempt to attract people. The natives understand the bird's behavior, and follow it to the hive. It is an unwritten law that the bird be allowed to feed upon the bee larvae.

The diet of late Palaeolithic humans would have varied from region to region, because by this time they were widely dispersed. The habitat of some humans was arctic, for others steppes, for others temperate woodland, for others savannah, for others tropical rain forest, and for others desert. A few of these habitats are occupied by hunter-gatherers even today, and in the next chapter we will be looking at some of these.

The economics of hunting and gathering must have influenced the biological evolution of human characteristics, as it has in other animals. Hunter-gatherer societies prevailed for at least a
million years of human history. Some experts consider that human life expectancy increased during this period[7]. Present day hunter-gatherers in Africa have a life expectancy that is similar to that of modern Europeans prior to the development of modern medicines[6], and this suggests that they are well adapted to their situation. In other words, evolution by natural selection has shaped humans as it has other animal species. Some have argued that humans are now free of such evolutionary pressures, but we shall see that the situation is complex, and the future unclear.

Now, what about you? How would you have fared 20000 years ago? Barring accidents, you could expect to live to the age of 60. You would have a healthy diet, one that your ancestors adapted to thousands of years ago. You body would be in tune with nature and your hunter-gatherer lifestyle would not be too onerous. One day you hear that some Neanderthals have trapped a female mammoth and it's calf in a gully. They have the animals corralled and are moving in for the kill. You rush over to watch the action. When you arrive you see that the Neanderthals have everything ready. Their women folk are assembled around a fire, with knives and scrapers, ready to deal with the carcasses. The men go for the adult mammoth, wounding it with their spears. Then they kill the calf, and they signal to you to take it over to the women. While they are attending to the adult mammoth, you carry the calf over to the fire, where the women start to work skinning and dismembering the body. They remove the stomach, fill it with water, and suspend it over the fire. Into the 'pot' they put the soft parts, heart, liver, etc. By the time the men have finished their work, the food is ready, and everybody joins in the feast, you included. After eating you depart, leaving the Neanderthals to deal with the adult carcass[7].

A week later you are starting to feel unwell, and you have strange marks on you body. You have the pox, and soon you will be dead. "I should have warned you", said an old woman in your community, "not to get too close to mammoths, because they carry a disease". But what about the Neanderthals? "They are immune, but we are not." This is our second case of *death by eating*[7].

Chapter 3 *The dangers of foraging*

Food may be gained in various ways, most of which can be listed under the heading of foraging. Exceptions include food that is donated to the young, to females during courtship rituals, and to domestic animals by their keepers. The food obtained by foraging is the animal's prime source of energy and specific nutrients. Food obtained by foraging includes that gained by searching, that gained by farming, and that gained by usurping (parasitism and stealing). Farming will be the subject of the next chapter, so let look first at searching for food.

Foraging efficiently

Natural selection favors efficient foragers, and animals are extremely adept at searching for and harvesting food. But foraging costs energy. Even sit-and-wait foragers expend energy while waiting, and active foragers have to spend energy to gain energy. So efficient foraging is a matter of energy economics.

Animals have to spend energy in order to forage and there may be circumstances in which the energy available to spend is limited. Similarly, the expenditure of a human shopper in a supermarket may be limited by the amount of money available, or by the amount of time available. The shopper with a limited amount of time is constrained by the rate at which items can be selected and paid for. The shopper with a limited amount of money is constrained in what they can afford to buy. The shopper has a time budget and/or a money budget.

So what is a *budget*? It is a limit to expenditure that constrains behavior. An animal cannot devote more energy to an activity than is has available. Nor can an animal devote more time than is available. When an animal has a choice between activity A and activity B, the consequences of both these activities entail the expenditure of energy and time. This limit on energy (or time) means that whatever is spent on A cannot be spent on B. If time spent of A or B (e.g. eating and drinking) were plotted in terms of the normal consequences of these activities (i.e. food and water obtained), we would see that the more time or energy is spent on A the less can be spent of B. The sum of the energy (or time) that can be spent would be expressed as a budget line on the graph. The same principles apply to humans, except that for many present day humans, the energy budget is replaced by a money budget. The analogy between energy and money is well founded. Consider the relationship between the price and the consumption of a commodity. People usually buy less of a commodity when the price rises, a phenomenon called elastic *demand*. However, the extent of the elasticity depends upon the availability of substitutes. For example, if the price of apples rises, then people tend to buy a lot less, because there are plenty of other fruit substitutes. When the price of coffee rises, people buy only a little less, because there are few available substitutes. Demand for apples is said to be elastic, and demand for coffee less so. Now, if an animal expends a certain amount of energy on a particular activity, such as foraging for a particular food, then it usually does less of that activity if the energy requirement is increased, especially when substitutes (other foods) are available. In fact, over the past 30 years it has been shown that many types of animal, when put to the test, obey the laws of consumer economics in very many respects. This is not all that surprising, once you realize that these micro-economic laws are basically a way of expressing the logic of choice. Any person or animal that has to expend energy (or money) to obtain necessary resources faces the same kind of situation, and it is not all that surprising that they all home in on the same (logical) laws governing economic behavior. Thus the elasticity of demand functions in animals gives an indication of the relative importance of the various activities in the animal's repertoire (such as feeding, grooming, territory defense, etc.).

Let us look at an animal example and a human example. Biologist Bernd Heinrich[1] studied foraging bumblebees , and likened them to a shopper.

"A bee starting to forage in a meadow with many different flowers faces a task not unlike that of an illiterate shopper pushing a cart down the aisle of a supermarket. Directly or indirectly, both try to get the most value for their money. Neither knows beforehand the precise contents of the packages on the shelf or in the meadow."

Heinrich's observations showed that time and energy act as constraints upon the bees' foraging efficiency. When the bees have to travel some distance to find productive flowers, then time becomes a limiting factor, and it is worthwhile for the bee to expend energy in order to save time. When foraging on relatively unproductive flowers or when foraging at a low temperature, the bees take more time in order to save energy. Similarly, a shopper tends to spend more money (per item) when in a hurry, but to spend more time when short of money.

Now consider the !Kung San, hunter gatherers that were extensively studied by anthropologist Richard Lee[2], before their way of life changed for ever. These people lived in the Dobe area of the Kalahari desert in Botswana. Lee found that individuals work to obtain food that is shared among all members of the group, plus any visitors. The consumption of each person is the same in relation to their requirements. In other words the food is shared out fairly. The San do not live a hand-to-mouth existence. They may eat a little of what they obtain in the field, but most is transported to the camp and shared out. Most of this food cannot be eaten right away, but must be processed. The hunting is done by males and the meat they bring home must be cut up, roasted, or hung out to dry. The females gather plant material that must also be prepared for eating. For example, Mongongo nuts must be cracked and roasted.

In terms of hours worked per week, the picture is as follows: The average man works 44.5 hours per week in total, earning just fewer than 500 kilocalories per hour. The time spent working (per week) is made up of 21.6 hours foraging, 7.5 hours repairing tools and weapons, and 15.4 hours work around the camp. The women earn 631 kilocalories per hour, working a 40.1 hour week. Their working time is made up of 12.6 hours foraging, 5.1 hours in repairs, and 22.4 hours work around the camp.

Whether foraging or doing camp chores, the individual is working for payment that is delayed. The payment comes in the form of shared-out food. If the payment had come in the form of money, it would have to have been spent on that same shared-out food, because there is nothing else to buy. If we assume that individuals are free to choose how much work they do each day, then we can put on an economist's hat (adopt an economic outlook) and anticipate a *trade-off* between work and leisure. In other words, the individual values work because of the money (energy) it brings in, and values leisure time because of the enjoyment of rest and social intercourse involved. So there comes a point (the *optimal* point) where the individual gains the same satisfaction from work done as from leisure time, and is indifferent between the two. If more time had been spent on work or leisure, then less satisfaction would have resulted. Of course, if the rate of pay for work were higher, or the returns of leisure-time greater, then the optimal point would shift. In fact, the optimal point shifts systematically as the wage rate increases, and the (mathematically described) shape of this shift is called the *labor supply curve.*

The point here is that every hour of work deprives the worker of one hour of leisure. The worker on a high income sacrifices a large amount of money to obtain increased leisure time, whereas the worker on a low income sacrifices little. This type of analysis is used by economists to predict how working people will react to changes in wages. It is devised for application to a monetary economy. But the !Kung San have no money. Nevertheless, by substituting energy for money we obtain the same results. In other words, the San divide their time between work and leisure in the same way as do people with money. Moreover, they work roughly the same hours per week as we do.[2]

In the 1960s, anthropologists woke up to the fact that many peoples with a hunter-gatherer lifestyle led a life that was very similar to modern 'civilized' people, in terms of their time spent working, and the quantity and quality of their food, and their enjoyment of leisure. Studies of the !Kung San in Botswana, of Australian Aboriginals, and the Hadza of East Africa showed that they rejected the farming lifestyle of neighboring peoples, and as a result they worked less and ate better[3].

In the 1970s zoologists and psychologists, working on animals, woke up to the fact that animals behave (economically) in a way that is very similar to economic behavior of people. (i.e. the same types of economic analysis can be, and have been, applied to animals as to people). Studies of numerous animal species have shown that they forage efficiently, and have plenty of time and energy for other activities[4].

In summary, in terms of energy (or money) the behavior of (hunter-gatherer) humans and other animals is very similar.

However, as we saw in Chapter 1, energy is not the only aspect of foraging that is important for survival.

Decisions decisions

Foraging efficiently implies that foraging should be profitable. However, in addition to deciding how and where to forage profitably, animals have other worries. The forager may have other dietary requirements apart from calories. It may be in competition with other individuals, the food on offer may be toxic, or there may be predators around. So the foraging animal has many decisions to make.

All animals depend upon certain resources, and all animals must be able to monitor three aspects of these vital resources. Firstly, the state of the resource within the animal's body must be measurable. That is, there must be information about the energy state and the state of other vital resources such as water (for land animals), specific nutrients, fat storage, etc. As we saw in Chapter 1, some of these factors are measured directly, and some indirectly. Secondly, the animal must be able to gauge the *availability* of the resource in the environment. For example, when you go to pick blackberries, you must be able to tell which bushes have blackberries and which bushes have none. A bush with many blackberries represents high availability, while a bush with few represents low availability. Thirdly, the animal must be able to gauge the *accessibility* of the resource in the environment. When foraging for blackberries you will find that some blackberries are difficult to reach. These berries may be numerous, but if they require a lot of time and effort to harvest, they have low accessibility. The rate of return for time spent harvesting (i.e. the speed with which you can fill your basket) depends upon both the availability and the accessibility of the blackberries. The berries may be numerous, but if you have to spend a lot of time walking between bushes, or climbing to reach the berries, then the rate-of-return will drop. There is not much one can do about the availability of a resource, but accessibility can often be improved by learning and by the use of tools. An experienced blackberry forager has learnt the skill of quickly picking a handful of berries, before depositing them in the basket. They may also have learned to bring along a walking stick to hook down high up branches with available berries.

The decision as to whether to forage for blackberries or do something else hinges partly on how much you want blackberries (your *motivational state* with respect to blackberries), and partly of your estimate of their likely *availability* and *accessibility*. If the indications are that there are few blackberries around, or that they are very inaccessible, because other people have picked the easy ones, then you may well decide not to go on a blackberry expedition. It is these three factors (state, availability and accessibility) in combination that influence your decision. It is much the same for other things that you might decide to do, and it is much the same for foraging animals.

In making a decision between mutually exclusive behavioral alternatives, such as sitting and standing, which cannot be performed simultaneously, there must be some kind of *common currency* (e.g. chalk and cheese could be compared in terms of common currencies like density or hardness). In economic terms the common currency is called *utility*. So while foraging for blackberries the question is whether you associate more utility with the situation in location A, or location B. Or, for that matter, whether you associate more utility with (or derive more pleasure from[5]) foraging for blackberries or watching TV. If it is raining, you may decide to watch TV.

The foraging animal faces a complex situation. First of all, should it go foraging at all? It may be hungry, but this must outweigh the importance of what it is doing before making the decision to go foraging. As we saw in Chapter 1, an incubating herring gull may find itself in a situation where it is becoming increasingly hungry, but to quit the nest to forage would be to invite egg predators. It turns out that the timing of the decision to quit the nest depends upon the amount of fat the bird is carrying. So the decision depends upon at least two internal variables – the animal's hunger and its fat reserves. In addition to information about its motivational state, the animal has to obtain information about the availability and accessibility of external resources. So the herring gull has to assess the situation as a whole, including the weather, the time of day, the state of the tide, what other gulls are doing, etc. The herring gull will have learned from experience to read the signs.

Many animals can learn to improve their foraging behavior, and this has been well studied by biologists[5.] For example, bumblebees with no field experience do not work on flowers as efficiently as do experienced individuals. They may land on inappropriate parts of the flower, or probe incorrect sites for nectar. With experience the bee soon learns to improve the accessibility of its food. In other words it is learning a skill. The bees also learn about the availability of nectar. Experienced bumblebees can learn which species of flowering plant is the best source of nectar, based upon the availability and accessibility of the nectar (i.e. nectar gained per unit time spent foraging) They can also learn to discriminate between constant and variable sources of nectar (The bees prefer the former). Once a bee has found a reliable type of flower, say a yellow flower, it tends to concentrate on that type of flower. However, if while foraging on yellow flowers, the bee passes blue flowers, it may sample a blue flower. If the overall energy profitability (energy gained – energy spent) is the same for yellow and blue flowers, then the bee will stick with yellow flowers, but if the profitability of blue flowers exceeds that of yellow flowers (perhaps because they are nearer), then the bee will learn to concentrate on blue flowers.

In addition to deciding how and where to forage profitably in terms of calories animals may have other requirements. John Goss-Custard[6] studied the foraging behavior of redshank a wading bird that hunts for food along the sea shore and on mudflats. He found that when the birds are feeding on worms they tend to pass over the smaller worms and select the larger ones. This size preference is influenced by the rate at which they encounter the larger worms but not by their encounter rate with small worms. This is what one would expect if the bird's foraging strategy was to maximize profitability, because the larger worms provide the greatest amount of energy per unit of energy expended on foraging. The smaller worms are not worth the energy expended in extracting them from the mud.

However, John Goss-Custard also found that when the amphipod crustacean *Corophium* was available in addition to worms, the birds tended to select *Corophium*. When he came to analyze the energy content of the prey, and the energy costs of obtaining the prey, he found that the birds would have obtained two to three times more energy (per minute) by taking only worms and ignoring the *Corophium*. Clearly, energy profitability was not the only factor influencing the redshank foraging behavior. Why did the birds decide to take the *Corophium* instead of the more profitable worms?

What might an economist have to say about this situation? To simplify matters we will refer to the redshank prey as worms and shrimps, and we will assume that each prey provides a certain amount of energy and a certain amount of some unknown nutrient. We know from Chapter 1 that certain specific nutrients are important ingredients of an animal's diet. We assume that the energy and nutrient are present in different proportions in the two types of prey. This means that the consequences of eating the two prey are different and a redshank with a certain energy budget can decide whether to spend on worms or shrimps, or on a mixture of the two (like a person deciding between apples and oranges in a supermarket).

The price of shrimps in energy terms is about twice that of worms (and suppose that the price of oranges is about twice that of apples), but the satisfaction gained from a worm and a shrimp (or an apple and an orange), of the same price, may not be the same. If I choose to spend my money on an orange rather than two apples, because the orange contains more sugar that two apples (say), then I am doing so because I gain more satisfaction, or **utility** as economists call it, by spending the same money on oranges rather than apples, even though two apples may contain more energy than one orange.

The utility of a commodity usually obeys the law of diminishing returns. Thus I may derive a certain amount of satisfaction or utility from my large china collection. If I add one more piece to my collection, then I will increase my satisfaction by a small amount. If I had a smaller collection, however, then adding the same piece would increase my satisfaction by a larger amount. So it may well be that the redshank grab the opportunity to buy shrimps rather than the more profitable worms, because they have little of some special nutrient that shrimps contain. Once they have feasted on shrimps, they may switch back to worms.

Rivals and enemies

When there is food around the hungry tend to gather. The foraging animal may find itself in competition with other individuals. This can have both positive and negative effects[7].

The former may come from gains in foraging efficiency. The fishing success of black-headed gulls increases with flock size up to about eight birds. A school of fish is more vulnerable if attacked simultaneously by several gulls. For lions and wolves, cooperative hunting by several individuals increases the success rate. The negative effects come from interference, and changes in the behavior of the prey. For example, oyster catchers feeding on mussels often have aggressive encounters over food. When mayflies are preyed upon by salmon they tend to congregate in 'safe' places, making themselves less available to the salmon. Moreover, a gathering of foraging animals is likely to attract scroungers and predators, as we shall see.

The behavior of other animals may provide cues to the availability of resources. For example, some birds in breeding colonies may rely on others to find food. In other words, some individuals (the producers) find resources, and other individuals (the scroungers) take advantage of the producers' finds. Among sparrows, for example, a scrounger may follow a producer that is searching for food, without doing any searching itself. When food is found, the scrounger may snatch it from the producer[9]. Sometimes the producers and scroungers belong to different species. Gulls are notorious for robbing other birds of their food. Gulls adopt vantage points amongst flocks of feeding lapwings. When a lapwing extracts a worm from the ground, it is immediately chased by the gull, and either drops the food, or takes flight with the worm in its bill. The gull gives chase in the air, and harasses the lapwing until it drops the food. Skuas are basically professional pirates. They pursue various species of marine bird and rob them of their prey. The victims usually regurgitate or drop their food, and the skua often catches the booty while still in the air. Skuas also steal the eggs and young of other species.

Not only do many foraging animals have to be wary of pirates, they also have to be wary of predators. A feeding animal is vulnerable. Most predators rely on an element of surprise to catch their prey, so consistent vigilance is a good insurance against predation. However, the solitary animal cannot afford to spend too much time watching out for predators, because some time must be spent feeding. Animals that feed in groups have the advantage that they can rely on the vigilance of other members of the group to some extent.

Biologist Brian Bertram[8] observing ostriches found that the feeding bird raises its head to scan for predators at random time intervals. This makes it impossible for a stalking lion to determine how much time it has to creep forward undetected, while the bird is feeding with its head down. Individual ostriches spend more time scanning for predators when alone than when in a group. The overall vigilance of the group (proportions of time that at least one bird has its head up) increases slightly as group size increases, so all the individuals benefit by feeding in a group.

The cost of vigilance is partly a matter of the time it takes, and this is offset by the benefit gained by detecting predators. When a predator is detected an alarm is given, but the individual giving the alarm may endanger itself by so doing. Such a risk could be avoided by doing less vigilance duty, but this would be a form of cheating[9]. Pigeons get around this problem. When feeding in a flock they give a special "intention" signal when they are about to depart. If they depart without giving this signal, then the other birds fly off in alarm. Thus the vigilant pigeon, by flying away without 'saying goodbye', is warning other members of the flock, but it gains an advantage by flying away first.

Human hunter-gatherers obviously face the same types of risks and dangers as other species. In Africa hunter-gatherers, such as the !Kung, Ache and Hadza are wary of lions, jaguars, buffalo, and snakes. The fear of predation must have some impact on their hunting tactics. However, other animals are not the only things that human foragers have to worry about.

Dangerous plants

We have seen that many plants protect themselves from animals by chemical means (Chapter 2). For example, Halogeton is a fast-growing annual plant of the Western USA.

Halogeton lacks the capacity to compete with vigorous perennial plants and the more aggressive annuals, but is a prolific seed producer. Wind, water, animals, and vehicles spread the seeds. Seeds may remain alive in soil 10 years or longer. Consequently, Halogeton often grows along railroad beds, roads, and sheep trails, and in places where the soil has been disturbed.

Halogeton frequently causes poisoning in sheep and cattle. It contains the toxic substance sodium oxalate, which is contained in leaves and other above-ground parts. Halogeton is dangerous at all times. It becomes more toxic as the growing season advances, reaching a peak of toxicity at maturity. The leaves have the highest concentrations and oxalate concentrations increase with maturity of the plant. The effect in poisoned animals is severe, about 1 gm oxalate/kg bodyweight is lethal in sheep. However, if fed at low doses for several days, the rumen bacteria become conditioned, resulting in increased tolerance. (lethal doses are about 30% higher). Sheep can tolerate large amounts of halogeton if they eat it slowly together with other forage.

Similar abilities to adapt to poisons occur in other species. In Alaska, deer eat a lot of skunk cabbage, a plant that also contains poisonous oxalates. The deer target skunk cabbage when it first emerges in the spring, eating the yellow flower spike and the green leaves. Although poisonous, it is rich in protein, critical to hungry deer after the lean pickings of winter. The way the deer deal with this situation is to eat a variety of plants, ingesting safely small quantities of toxic plants and focusing on higher nutritional quality plants when they are available. Deer will eat more than one type of vegetation in a meal, and the combination of plants in the digestive tract at the same time may minimize or diminish the toxic effects of some foods. Researchers have found that animals will eat clay or lick minerals from the soil that tend to buffer or bind to toxins and counteract the harmful effects of plant compounds.

Herbivores tend to be wary of plants that are new to them. David Mackenzie[10] has a good description of the behavior of goats towards novel food.

"Safety from poisoning in the natural flock is achieved by the discipline of the foraging expedition, which demands that every goat eats the same kind of herbage, in the same patch and at the same time as the leader of the flock........The first goat to spot an unfamiliar food, stops as if shot, snorts, and adopts a tense stance with feet set wide and ears pricked, ...Then comes a cautious inquisitive approach, nose stretched forward and the body bunched for lighting retreat. Then the flock queen takes over the investigation, adopting the same poses, and the rest of the flock huddle and watch. First the flock queen noses the object; then she nibbles a bit and spits it out; nibbles again and chews tentatively; if good, she swallows it, and, as if revelation had suddenly come upon her, proceeds to wolf it. The other members of the flock follow her example, If the mysterious food is not good, the flock queen spits it out with vehemence, snorts, fusses, and frantically runs about wiping her mouth on the grass and pawing at it......"

We can imagine that human hunter-gatherers have a similar kind of foraging discipline. You do not pick or eat a plant that you do not recognize, without first consulting the most experienced member of the group. You do not venture onto new ground without first consulting the boss. In many ways the earlier hunter-gatherers were similar to foraging animals. They faced the same issues and dangers, yet became adept at making a living out of their particular environment. Through the process of natural selection, they inherited traits that enabled them to adapt to the environment, and they were also able to learn for themselves, and learn from their community. The only fundamental difference between humans and other animals is that humans evolved in such a way that they had greatly enhanced access to shared knowledge, culture and tradition.

Anthropologists[11] are well aware that present day hunter-gatherers have in intimate knowledge of all the plants and animals around them, including plants that require special preparation to make them non-poisonous and plants that have medical uses. So we can be fairly confident in supposing that Paleolithic humans were also sophisticated in a similar way, they would have learned from experience. They had accomplished a 'Great Leap Forward' (Chapter 1), equipped themselves with tools made from wood, bone and stone, and invented clothes and domestic fire. These accomplishments enabled them to populate Africa, Eurasia, the Americas and Australia[12] Some of their communities have survived to this day.

Now, what about you? Imagine that you were a human hunter-gatherer 11.000 years ago. One day you set out, together with other women and children, on a plant foraging expedition. You go to a place where there are wild onions, and you instruct the children in the art of digging these up without breaking the bulbs. That evening you walk back to the camp and start preparing the evening meal. The onions are roasted in some fire embers, and eaten with some game that the men have managed to bag. The next morning you are ill. You have abdominal pains, nausea, light-headedness, shivering, and convulsions. Within a week you are dead. You have been unlucky. One of the children dug up narcissus bulbs, mistaking them for onions. These are poisonous even when cooked[13]. If you had eaten only one, you would have been ill, but you ate too many. If someone in the group had noticed the difference between narcissus and onion bulbs, you would have been forewarned. But in their enthusiasm, the children piled the bulbs onto the fire, before they were properly inspected. This is the way that lessons are learned and experienced gained. This is our third case of *death by eating*.

Chapter 4   *Growing your own*

The diet of Palaeolithic and Mesolithic communities (early and middle Stone Age) is often held up as a model for present day peoples. The idea is that, because our species evolved for a million years as hunter-gatherers, our bodies are attuned to a hunter-gatherer diet, and not to a modern diet[1]. In other words, we are genetically similar to Stone Age Man.

There may be some truth in thus idea, but we have to remember that during the past 50000 years humans moved out of tropical Africa, and spread around the globe. Australia, which is largely desert, has been inhabited for 40,000 years. Siberia, which is very cold, has been inhabited for 20,000 years, and northern Europe for 40,000 years. During this period there was an ice age (from 80 to 15 thousand years ago); producing glacial periods during which many humans had to survive in conditions very different from those of tropical Africa. In colonizing these alien environments, humans must have come under intensive pressures of natural selection. In fact we know that there were different genetic changes in the various groups of humans, as a result of DNA studies[1], and because we can tell which part of the world people come from simply by looking at them. Many of these changes have to do with climate, but some have to do with diet, as we shall discover.

It is true that hunter-gatherers, whatever part of the world they inhabit, did not eat some of the things that we eat today. During the past 10,000 years our food, part of our **extended phenotype**, has changed dramatically. The results of this change are complex, but before we delve into these, let us see why and how this change occurred.

The Neolithic Revolution.

The term Neolithic Revolution was coined by Gordon Childe[2]. By this term he did not intend to imply the kind of revolution that we think of in historical terms. He was referring to a revolution in the human way of life, from a nomadic existence to a settled existence that became possible with the domestication of plants and animals. This revolution has occurred among different peoples at different times, and even today there exist people who have not made this transition.

As an example let us look at the domestication of reindeer, the only kind of deer to be domesticated until recently. Reindeer are indigenous to northern Eurasia, and North America (where they are known as Caribou), a land that was covered by ice during the last glacial period. Reindeer were once found as far south as Nevada and Tennessee in North America and Spain in Europe, but the ice started to recede about 12,000 years ago. Today, wild reindeer have disappeared from many areas within this large historical range, especially from the southern parts where it vanished almost everywhere. Large populations of wild reindeer are still found in Siberia, Greenland, Alaska and Canada. The last remaining wild reindeer in Europe are found in portions of southern Norway. Domesticated reindeer are mostly found in northern Scandinavia, Russia and Iceland. The southern boundary of the species' natural range is approximately at 62° north latitude, though this may change as a result of global warming.

The great value of this species lies in its exploitation of the tundra. Reindeer are ruminants, eating mainly lichens in winter, especially 'reindeer moss'. However, they also eat the leaves of willows and birches, as well as sedges and grasses. Reindeer hunting by humans has a very long history and it is probable that humans started hunting reindeer in the Mesolithic and Neolithic periods. Norway and Greenland have unbroken traditions of hunting wild reindeer from the ice age until the present day. In the non-forested mountains of central Norway, it is still possible to find remains of stone built trapping pits and guiding walls built especially for hunting reindeer, probably during the Stone Age. Wild caribou are still hunted by the indigenous people of North America and Greenland.

Reindeer have been herded for centuries by several Arctic and Sub-arctic people including the Sami (Lapps).They are raised for their meat, hides, antlers, and also for milk and transportation. Reindeer are not considered fully domesticated, as they generally roam free on pasture grounds. In traditional nomadic herding reindeer herders migrate with their herds between coast and inland areas according to an annual migration route and herds are keenly tended. However, reindeer have never been bred in captivity, though they have long been tamed for milking as well as for use as draught animals and beasts of burden.

The indigenous peoples of North America, such as the Inuit (Eskimos) hunted caribou, but did not herd tame them for milking. Even faced with the same species in a similar environment as North Eurasian peoples, they did not bring these animals into semi-domestication. As we shall see, this difference is probably due to genetic differences between these two peoples of differing origin. Thus whereas the Inuit diet is composed of marine mammals, fish, caribou, small game, birds, and plants, the Sami depend on herded reindeer for milk and meat, and on fishing, gathering plants, and hunting small game and birds. Thus, in these northern extreme conditions we see (until recently) humans on the verge of the Neolithic revolution.

In warmer parts of the world, domestication of animals and plants began much earlier. Dogs were domesticated about 17000 years ago, and these would have been useful as guard dogs, and as an aid to hunters. Sheep and goats were domesticated about 12500 years ago[3]. It is worth noting that these animals, dogs, sheep and goats, could have been kept by nomadic peoples, herded along when the group was on the move. Pigs were domesticated somewhat later, about 11000 years ago, but these would be kept by people in settled villages. Similarly, cattle were first domesticated about 10000 years ago.

Domestication of plants began at different times in different places. For example, wild yams (Dioscorea) were gathered by hunter-gatherers, but it was usually necessary to remove their toxic alkaloids by boiling. Domestication of yams developed independently in tropical America, Southeast Asia and West Africa [3]. Archeological evidence shows that capsicum peppers were eaten 9000 years ago, in the Americas but these were probably wild plants. Domestication of peppers is thought to have started 7000 years ago, in two separate locations. Phaseolus beans were native to the Americas, and archaeological evidence indicates that were cultivated some 8-9000 years ago. A similar story can be told about groundnuts. However, the Neolithic revolution is usually exemplified by the domestication of wheat and barley in Egypt.

About 11000 years ago there were established villages in the Eastern Mediterranean that relied on a combination of hunting and intensive gathering. Wild grasses in that area had edible seeds, and a man with a flint-bladed sickle could gather two pounds of grain in an hour[4]. The inhabitants probably discovered that by leaving some plants to disperse their seeds, they were more likely to have plants to harvest the next year. It is a small step to plant some seeds yourself. Cultivation of crops from this area (often called the Fertile Crescent) spread rapidly throughout the Mediterranean, reaching central Europe in two thousand years, and Britain and Scandinavia in another two thousand years.

The changing lifestyle

Early Neolithic people went to a great deal of trouble to prepare wheat and barley. The edible part of the grain (consisting of the embryo and the endosperm) had to be separated from its protective coating (the bran and the chaff). Archaeological investigations, in Egypt, Syria, Iraq and the Jordon valley, show that this was done by heating the grain on hot stones

Cooking became easier with the invention of pottery. But pottery is not of much use to nomadic people. It is too heavy to carry around, and it breaks easily. The first pottery cooking vessels were made in Japan 14000 years ago. By this time some Japanese people had settled in fishing villages on the coast, where they exploited the marine resources[5]. Pottery was used for cooking purposes in the near East 8000 years ago, and in the Americas by 3500 years ago[6].

A big advantage of grain is that it can be stored, but a disadvantage is that such stores cannot be easily moved. Therefore, communities that cultivated grain had to secure *territory* on which they could live and grow and store their crops. The inhabitants of such settlements would find it more difficult to go hunting, because they would soon exhaust the game supply in their area. Unlike, nomadic people, they could not follow migrating animals, or move to new hunting areas. To have a good supply of meat, they would have to keep animals in semi-captivity, and they would have to tame them. In other words, they would have to invest in animal domestication. For these, and other, reasons, their diet was bound to change.

The changes in human diet during the Neolithic period would vary from region to region. Those growing crops inland would eat much more carbohydrate than their Paleolithic ancestors, those keeping reindeer, goats, or cows would eat more dairy produce, while those living on the coast would have a relatively smaller change in their diet. Human food is a result of human behavior, part of the human extended phenotype (see Chapter 1), and there is no doubt that there was a large evolutionary change in human food as a result of the Neolithic revolution. This change had three possible and important effects: (a) It changed humans genetically. (b) It changed the environment and (c) It changed the human lifestyle, with consequences for health and reproduction.

Taking (a) first, it is often thought that 10,000 years is not sufficient for animals to be changed genetically, and for those changes to become established. When we look at domestic animals, we can see that this is clearly untrue. Many domestic animals have changed radically from their wild ancestors. What is true is that the rapidity of evolutionary change depends partly upon the generation time, and upon the selective pressures applied. For example, a female dog can have puppies every two years, and the human dog breeder can exert extreme selective pressure by preventing dogs that do not show certain characteristics from breeding. Charles Darwin was well aware of this kind of *artificial selection*. It might be argued that humans have a much longer generation time than dogs (true), and that there is no alternative to natural selection (false). The alternative to natural selection is *sexual selection*, 'the advantage which certain individuals have over others of the same sex, and species, solely in respect of reproduction'[7]. For example, in the Paleolithic period, all humans had brown eyes. Blue eyes appeared about 10,000 years ago, as a result of a mutation[8]. If women preferred men with blue eyes, then those men would tend to father more children than men with brown eyes. So, in biological terms, men with blue eyes would have greater reproductive success, and because eye color is inherited, their blue eyed children would also be at a reproductive advantage. Now brown eyes contain pigments that protect the eyes from sunlight, so one would think that brown-eyed people would have an advantage due to natural selection. Overall, this is probably true in tropical areas were the sunlight is strong, but in northern climes the sun is not so strong. So it is entirely possible that sexual selection wins out in some parts of the world, but is overpowered by natural selection in others.

Let us now look at another well-studied example. The normal mammalian condition is for the young of a species to exhibit reduced production of the enzyme lactase at the end of the weaning period. In non dairy consuming human societies, lactase production usually drops by about 90% during the first four years of life. People lacking lactase cannot digest milk. However, certain human populations have an inherited mutation[9] that results in a bypass of the shutdown in lactase production, making it possible for members of these populations to continue consumption of fresh milk and other dairy products throughout their lives. If your remote ancestors were pastoralists, keeping cattle and goats, then you are likely to be lactose-tolerant and able to digest milt as an adult. Scientists believe that Northern Europeans and certain African tribes evolved lactose tolerance independently, whereas people of Mongolian descent do not have the genetic makeup that enables them to digest milk as adults[9]. As we have seen (above), reindeer herders of Mongolian descent have no milk in their diet, whereas those of European descent do make use of the reindeer milk.

Genetically, we are very similar to the hunter-gatherers of the late Paleolithic, but this does not mean that our evolution has not continued during the past 10,000 years. The evidence[10] suggests the opposite. However, the pace of genetic change is outstripped by the circumstances on the ground. There were big changes in human diet as a result of the Neolithic revolution, and there were big changes in the environment (see below). This has let some scholars to suggest that we humans now suffer from *evolutionary discordance* – a mismatch between the situation that our bodies have evolved to respond to and the situation that we face now. Our genetic makeup reflects an evolutionary compromise between multiple, competing selective pressures. Our physiology is therefore a reflection of adaptations that must function simultaneously in concert, and that mutually and intimately affect one another. (This is sometimes called *co-adaptation*). The discordance hypothesis, attempts to assess the disjunction between the ancestral environments and the ones we live in now, and to predict points of vulnerability due to the rapidity of environmental change. In particular, several important chronic degenerative diseases have been interpreted as "diseases of civilization" because they appear to result from this disjunction[11]. As we shall see (Chapters 9 and 10) a number of dietary complications and diseases are thought to be due to this phenomenon.

Turning now to the changing environment during the past 10.000 years; A major impact of agriculture has been deforestation. Early Neolithic farmers would clear land for cultivation, using a slash and burn approach, and this practice carries on to this day. Wood was the main source of fuel for thousands of years, and as the human population grew, their appetite for wood grew. Trees were felled to provide wood for fires, buildings and ships. The vast woodlands and forests that appeared at the end of the last ice age are now severely attenuated. Not only has this changed the landscape in many places, but it is also thought to have had an effect upon the climate[11].

Agriculture also had other effects upon the environment. In Mesopotamia over the past 6000 years, the development of irrigation systems has led to a considerable increase in the agricultural area, but since the fourth millennium, large area losses due to irrigation mistakes (salinization) have occurred. Deforestation and erosion processes are known to have originated in several regions of the Mediterranean area over the past 7000 years[12].

How did these environmental changes affect mankind? We have seen (Chapter 2) that the life expectancy of Paleolithic hunter gatherers was not much different from that of 18th century Europeans. *Life expectancy* is the average number of years a human has before death, conventionally calculated from the time of birth (Note that this measure includes infant deaths) For Neanderthal Man the life expectancy was 20 years, and for Paleolithic hunter gatherers it was 33 years. So *Homo sapiens* was doing better than *Homo neanderthalensis* , and the latter became extinct. During the Neolithic period, however, human life expectancy dropped significantly (20 years). In classical Greece and Rome it was 20-30 years, but in Pre-Columbian North America it was 25-35 years. In Medieval Britain life expectancy was 20-30 years, and in the early 20th century it was 30-40 years. Now, the current world average is 67 years, and rising[13].

So we can see that the Neolithic revolution was accompanied by a marked drop in life expectancy. Of course, the change to an agricultural way of life occurred in different places at different times, so the question is – did the same effect occur at these different times. The answer is yes. Wherever and whenever people changed from a hunter-gatherer to an agricultural way of life, life expectancy declined[14]. The evidence suggests that farmers suffered higher rates of infection due to the increase in the size and permanence of human settlements, poorer nutrition due to reduced meat intake, and greater interference with mineral absorption by the cereal-based diet. Consequently, Neolithic farmers were shorter and had a lower life expectancy relative to their hunter-gatherer ancestors. However there was a rise in the birth rate and large population increase.

Studies, by Richard Lee of the Kalahari !Kung living a hunter-gatherer existence show that women are severely encumbered by their children. The women provide about two-thirds of calorie income by foraging for plant foods. Mongongo nuts are their most important food source. These are in plentiful supply in the dry season, but are usually situated some six miles from suitable camp sites. The women make excursions to gather these nuts every few days, taking their children with them. The men take no part in plant gathering, but confine themselves to hunting. Lee shows that the weight carried by mothers on these excursions increases with the frequency of having babies, not only because there are more mouths to feed, but also because the small children have to be carried. Nick Blurton-Jones and Richard Sibly show that an average birth spacing of four years is the optimal under the prevailing conditions. Thus the women maximize their reproductive success by spacing births widely and by foraging seldom. The women have children more frequently when they do not have to carry them on foraging expeditions[15].

In other words, in hunter-gatherer societies, women need a gap of at least three to four years between children, because highly dependent babies are incompatible with a mobile lifestyle. No such limitation exists when people live in permanent settlements, and so, as a result of the Neolithic revolution, it became possible for women to have children more frequently. As the techniques of plant cultivation and animal husbandry became more refined, it was possible to feed larger groups of people from relatively small numbers of food-sources, and still have food left over for storage during the winter months. The ability to settle in one place, and to store food, led to a population explosion.

As Mark Cohen has remarked, "Slightly more than 10000 years ago, the overwhelming majority of people lived by hunting and gathering. By 2000 years ago, the overwhelming majority of people lived by farming". The human population 2000 years ago was much larger than it was 10,000 years ago, so agricultural practices must have had a big impact of the environment in general (see above) and on the lives of individuals, in particular. In towns and villages, the population density was high, and the risk of cross-infection increased. Children of agriculturalists, especially, were more likely to die than were their hunter-gatherer counterparts. Land became enclosed for planting purposes, and people demanded property rights for the first time in history. As we shall see in the next chapter, this led to political developments, and there was much less equality than there had been in hunter-gatherer societies. A higher population density increases the level of technological sophistication, which in turn increases total productivity, allowing for further increases in population density[16]. There comes a point where the environment reaches its *carrying capacity*.

In biological terms, the *carrying capacity* of the environment is the equilibrium population size of a species that can be supported by a region. In anthropology, it has been defined as the maximum human population that the region can support without progressive degradation[17]. These definitions are somewhat simple, ignoring the possibility that there may be no static equilibrium, because the population can cycle or even vary chaotically. Moreover, the environment may be modified by the species to increase or reduce the population that it can support. In other words, the carrying capacity is partly a result of the environment itself, and partly a result of the behavior of the inhabitants, whether they be humans or other animals. For example, if goats are allowed to forage free and unrestricted, they soon exhaust their food supply, especially in dry areas. So an area of land that could support many goats turns into an area that can support fewer[17]. Technological attempts to alter the carrying capacity of the environment can be a double-edged sword, as we shall see later.

The dietary consequences

The profound changes in lifestyle that began with the introduction of agriculture and animal husbandry 10000 years ago occurred too recently on an evolutionary time scale for the human genome to adjust. This *evolutionary discordance* has been claimed by some to be responsible for modern diseases of civilization. In other words, the difference between our ancient, genetically determined biology and the nutritional, cultural, and activity patterns of Neolithic agricultural peoples, is a change too far. In particular, food staples and food-processing procedures introduced during the Neolithic period have fundamentally altered various nutritional characteristics of the ancestral human diet.

However, as we have seen (above) some genetic adjustments have taken place during the past 10000 years, but these have not become universal human characteristics. Therefore, the inhabitants of some areas (e.g. pastoral peoples) would have made some genetic adjustments, and the inhabitants of other areas may have made other adjustments. Increasing urbanization, and frequent migrations have led to a mix of people with differing genetic adjustments, and this has led to all sorts of further complications. Cereals, beans, animal fat and salt, in particular, can cause problems for people who are not genetically adjusted to these foods.

Because wild cereal grains are usually small, difficult to harvest, and minimally digestible without grinding and cooking, the appearance of stone processing tools in the archaeological record represents a reliable indication of when and where cultures systematically began to include cereal grains in their diet. Ground stone mortars first appeared long ago, used by people foraging for wild plants, but domestication of various types of wheat by their descendants heralded the beginnings of early agriculture and occurred by about 10000 years ago from strains of wild wheat localized to south eastern Turkey. Cereal grains were rarely consumed as year round staples by most worldwide hunter-gatherers, except by certain groups living in arid and marginal environments.

For some Neolithic societies, fifty percent of their calorie intake came from cereals. This diet would be supplemented with various fruits and vegetables, and occasional meat, but the amount of carbohydrate is much larger than that of their hunter-gatherer ancestors. Nowadays, we can estimate the *glycemic index* of various types of cereal foods, such as wheat, rice, etc. and what we find it that refined rice and wheat and potatoes have a relatively high glycemic index compared with whole grains, pulses, and sweet potatoes. The glycemic index is an index of the blood glucose raising potential of food items. The *glycemic load* is a measure of the blood glucose raising potential of a meal of a standard size. So although early Neolithic peoples probably ate unrefined (low glycemic index) grains,etc., but they ate a lot of it. Consequently they would have a high *glycemic load* compared with their non-farming ancestors. A high glycemic load is linked to diabetes[18].

Beans were also cultivated by early agriculturalists, but many species of beans contain dangerous toxins. For example, a substance in fava beans triggers the hemolytic anemia in some genetically susceptible individuals[19]. Because only some members of the population are affected in this way, it would have been difficult for members of the community to make the connection between their illness and the consumption of beans. Similar problems can be associated with the consumption of wheat.

Before the Neolithic period, all animal foods consumed by humans were derived from wild animals. The absolute quantity of fat in wild mammals is dependent on the species, but (apart from marine mammals) the wild animal carries much less fat than its domesticated counterpart. Moreover the type of fat found in wild animals is less deleterious for the human diet[20]. We have seen that a diet high in milk products is common in many parts of the world, and although some people not able to digest milk as adults, they are able to consume dairy products such as yogurt, butter and cheese, all of which contain saturated fatty acids. A high content of saturated fat in the diet is linked to vascular problems that may result is stroke or heart attack, and these diseases are common amongst people who eat relatively large amounts of (domestic) red meat and dairy products.

There is evidence from China 8000 years ago, and Europe 6000 years ago of systematic mining, manufacture, and transportation of salt. It is probable that Paleolithic hunter-gatherers living in coastal areas dipped food in seawater or used dried seawater salt, but on the whole hunter-gatherers did not add salt to their food[21]. Furthermore, there is no evidence that Paleolithic people undertook salt extraction or took interest in inland salt deposits. Amongst settled agricultural peoples, salt was used to preserve meat during the northern winter, and it was later regarded as an important flavor enhancer. As we saw in Chapter 1, excess salt is dangerous.

In summary, the Neolithic revolution, started in different places at different times, and in some places it never started, the inhabitants remaining hunter-gatherers until very recently. The revolution brought about a large increase in the human population and reproductive success, but the lifespan of most individuals was reduced. This reduction was partly due to excessive consumption of foods to which the people were not evolutionarily adjusted, and partly to changes in lifestyle, such as living in one place. Nomadic peoples move on, and to some extent leave their enemies behind. The longer you stay in one place the more you attract rats and mice, and other disease-bearing creatures. The more you stay in one place, the more the population increases, and the greater the chance of cross-infection. The advantages of staying in one place are that pottery can be made and used for more efficient cooking. Food can be stored against lean times, and various community activities can develop. Unfortunately, some of these activities are not all that desirable, as we shall see.

So how would you have fared during the Neolithic revolution? You would have lived a settled agricultural existence, and your diet would be high in carbohydrate. You live in a village, where you help to gather grass seeds and home-grown grains. You help to thresh them, dry them in the sun, and store them. One winter's day you are asked to fetch some grain from the store – it is moldy – use it anyway, we have no other. You use the grain to make flatbread. It will be OK, the fire will destroy the mould.

72

Two weeks later, YOU are dead. This is our fourth case of *death by eating*. Some of the fungi that grow on stored cereals produce toxic substances that are not destroyed by cooking. [22]

Chapter 5
The risks and benefits of exchange

Amongst animals, food is disposed of in one of five ways: It can be lost, hoarded, eaten, donated or sold. Food may be lost to competitors or robbers (see Chapter 3). It can be eaten on the spot or taken to another place, where is may be hoarded, or donated to a mate or to the young. Finally, amongst humans, food may be sold.[1]

Amongst humans we find subsistence living of various different types. Historically, the most primitive was regarded as the hunter-gatherer existence, based on wild plants and animals. As we have seen, some authorities think differently these days (see Chapter 4). Primitive agriculture leads to domestication of plants and animals, and to more complex economic organization. Except when working individuals consume the fruits of their own labors the products of human labor are distributed by means of exchange, the practice of giving and receiving valuable objects and services. Most hunter-gatherers practiced some form of reciprocal exchange, the giving of goods and services that is not contingent on any definite receipt of goods and services in return. This practice persists to this day, in our societies, at Christmas and other festivities. The giving is based on a general understanding that there will be some eventual reciprocity.

Reciprocal exchange is open to invasion by cheats, or free-loaders. They receive favors from others, but do not reciprocate. In some cases, such as the giving of food or clothing to children, no reciprocation is expected, but a roughly symmetrical reciprocity is expected amongst adults. In societies practicing reciprocal exchange free-loaders are subject to subtle form of disapproval, but demands are not made upon them in particular cases. It is usually part of the ethos of reciprocal exchange to deny that any balance is being calculated. Those that provide do not boast, or expect to receive thanks, as the anthropologist Marvin Harris explains,

"Boastfulness and acknowledgement of generosity is incompatible with the basic etiquette of reciprocal exchanges. Among the Semai of Central Malaya, no one even says 'thank you' for the meat received . . . Having struggled all day to lug the carcass of a pig home through the jungle heat, the hunter allows his prize to be cut up into exactly equal portions, which are then given away to the entire group . . . to express gratitude for the portion received indicated that you are the kind of person who calculates how much you are giving and taking . . . To call attention to one's generosity is to indicate that others are in debt to you and that you expect them to repay you. It is repugnant to egalitarian peoples even to suggest that they have been treated generously."[2]

Reciprocal exchange is more likely to occur amongst people who are genetically related, partly because hunter-gatherer groups are usually inter-related, and partly because it is human nature to favor your kin. There is scientific evidence showing that the degree of altruism towards kin is a function of kinship distance (how close the genetic relationship is), and there is anthropological evidence indicating that the extent of reciprocal exchange is in fact related to kinship distance in hunter-gatherer societies[3].

Reciprocal altruism also exists in animals, where it provides scope for cheating, in that an individual that receives may refuse to reciprocate at a later date. For natural selection to favor reciprocal altruism, the individuals must have suitable opportunities for reciprocation, and must be able to recognize each other individually and remember their obligations, and they must be motivated to reciprocate. These conditions are found in some human hunter-gatherer societies, and it has been suggested that reciprocal altruism has played an important role in human evolution[4].

However, once people become settled, and community size increases, then simple reciprocity breaks down. Reciprocal exchange works well when productivity is not over-encouraged, and the supply of food remains stable. If the human population increases, more food is required, and individual providers tend to be praised and encouraged. Successful providers acquire a reputation, and their status in the community changes.

It is noteworthy that many hunter-gatherer communities under-exploit their food resources. For example, "Although Hazda, in common with probably all other human societies, do not eat all the types of animal available to them – they reject civet, monitor lizard, snake, terrapin, among others – they do eat an unusually wide range of animals. ... In spite of the large number of species which they are both able to hunt and regard as edible, the Hazda do not kill very many animals and it is probable that even in the radically reduced area they occupied in 1960 more animals could have been killed of every species without endangering the survival of any species in question."[5]

Exchange and politics

It is a short step from simple egalitarian reciprocal exchange to egalitarian redistribution. Imagine a community where some members go hunting, others gather wild plants, and others cultivate the land. Then they all get together and pool their resources. Instead of a producer sharing what has been gained that day, the producers now give their gains to a central pool, and somebody then distributes the food throughout the community. That somebody has a special status and responsibility in deciding who gets what. Politics has now entered the arena.

Of course, pooling probably always existed within the family, because children do not help themselves, but are assisted or supervised by an adult. But in a wider arena pooling involves a shift in the social order. This shift has been a major topic of discussion amongst anthropologists[6], and it is clear that the repercussions have been different in different communities. However, there is one effect that is of importance to us, and that is the increase in productivity that is heralded by this type of change in social organization. Once a person takes responsibility for distributing food, that person is not only marked out as different, that person has an incentive to encourage productivity. When things are bad, the focus is on the distributor. When things are good, that person acquires a halo. So there gradually develops a political vested interest in increasing productivity.

To quote Marshal Sahlins "Agriculture not only raised society above the distribution of natural food resources, it allowed Neolithic communities to maintain a high degree of social order where the requirements of human existence were absent from the natural order. Enough food could be harvested in some seasons to sustain the people while no food would grow at all; the consequent stability of social life was critical for its material enlargement. Culture went on then from triumph to triumph, in a kind of progressive contravention of the biological law of the minimum, until it proved it could support human life in outer space – where even gravity and oxygen were naturally lacking."[6]

In other words, while nomadic Paleolithic hunter gatherers could move on when things got bad, their settled Neolithic descendents had to devise ways of storing food, and distributing stored food. If the special person who has the responsibility of distributing pooled food, is not themselves a food producer, then we have the beginnings of what Jared Diamond calls a kleptocracy[6].

"chiefdoms introduced the dilemma fundamental to all centrally governed , non egalitarian societies. At best, they do good by providing expensive services impossible to contract for on an individual basis. At worst, they function unabashedly as kleptocrats, transferring net wealth from commoners to upper classes. These noble and selfish functions are inextricably linked,."

Once the supply of food is under political control, things start to change. Firstly, accounts are kept. In a system where productivity is encouraged, there has to be some kind of account of who produced what. Agricultural people can often benefit from increasing production, provided they do not endanger their food supplies. They can afford to admire and encourage those who are big providers, provided things do not get out of control. To control productivity some kind of policing is required. Whoever produces must contribute, but someone must also safeguard the means of production. This kind of situation leads to stratified redistribution, with the controllers at the top, the producers in the middle, and the consumers at the bottom.

Exchange amongst unrelated (non-kin) groups usually takes place at some kind of market place. In the absence of money, goods and services are bartered. There is bargaining between the opposing parties, each trying to maximize their profitability. If the barter exchanges are not direct, then some sort of accounts must be kept. There must be mutual trust, or some kind of policing. This type of barter market can work satisfactorily if relatively few commodities are involved in the exchange.

A society that has some form of all-purpose money can operate a price market. The producers sell their goods in exchange for money, and use money to purchase other goods or services.[7]The main advantage of a price market is that it can handle many commodities in a mixed economy. Another advantage is that it facilitates trade with strangers, with whom no elaborate trust arrangements exist. Traditionally, money has been regarded as a portable, recognizable material. It has a certain legality that makes it divisible and convertible, and confers wide generality of use. In recent times, however, we have started to move away from money as a physical object of exchange. Bank accounts and credit cards enable us to promise each other that we will pay our debts at some time in the future. In this respect we seem to be returning to some features of reciprocal exchange.

What happened to the food?

We are interested in the evolution of human food, food a product of human behavior. As human behavior changes over time, so food changes. Food is part of the human *extended phenotype*. During the Neolithic revolution, humans began to domesticate plants and animals, albeit at different times in different places. So how did our food change?

As we saw in Chapter 4, domestication led to (some) humans consuming milk as adults for the first time; and to eating domesticated, rather than wild; animals, and to there being a greater proportion of carbohydrate in the diet. As the nutritionist John Yudkin has pointed out, the Paleolithic human was omnivorous, and "his diet when hunting was good was relatively rich in protein, reasonably rich in fat, and not very rich in carbohydrate….it seems reasonable to suppose that the

79

particular ecological niche in which man found himself 9000 years ago was one to which he had become pretty well adapted, and that adaptation included foods that he chose to eat."[8]

However, once the individual is no longer in control of the supply of food, the choice starts to become limited. Hunter-gatherers can pick and chose from the day's haul, and can respond to any *specific hungers* (see Chapter 1) that arise, except on a bad day, when the haul is meager. In the long run, however, they have a diet that enabled then to adjust to their bodily needs. Once the availability of food comes under political control, much dietary flexibility is likely to be lost. Even today, there are many communities that have a restricted and monotonous diet.

Political control over food distribution leads to rationing. During the Second World War rationing was introduced in the UK to try to ensure that food was distributed fairly. But there was also unofficial rationing, because it was too dangerous to ship in fruit from foreign countries, and people had to do without certain items that they had become used to eating before the war, such as oranges and bananas.

For early Neolithic communities, some of the foods that their ancestors enjoyed became less available. This happened partly because they were no longer nomadic. Some of the game and wild plants in the surrounding to their settlements would quickly become exhausted. Not only would there be a decline in availability, but also a decline in knowledge. No longer would everybody learn to identify those plants that were useful for food or medicines. No longer would everybody learn how to catch small animals. Those working in the fields, or in towns, had reduced opportunity to make such discoveries, and the previously widespread knowledge of nature gradually became the domain of specialists. Knowledge is power, and some specialists would gain privileged status.

Some settled communities became excessively dependent upon particular staple foods, such as yams in Africa, corn in Middle America, millet and rice in China, and cereals in the Mediterranean. Many cereal crops are low in protein, but that was usually made up by cultivation of pulses. Thus groundnuts in Africa, beans in Middle America, soybean and mung bean in China, and peas and lentils in the Mediterranean, all became important providers of protein. Nevertheless, there was malnutrition, especially in stratified societies where the poor had to eat what they were given, while the rich ate what they chose. In Roman times, the basic diet of the poor "consisted of grain-pastes or coarse bread bristling with chaff, and a polenta-like porridge made from millet. Water was the usual drink. Cooking was primitive, because equipment was primitive, fuel a problem, and fire a high risk.... As a result, the poor avoided cooking, whenever possible."[9] For the Roman rich it was different. The Roman Empire imported vast quantities of food from all over the known world, spices from Indonesia, pomegranates from Libya, pickles from Spain, and oysters from Britain.[9] How was this possible? The answer lies in trade.

Trade

Trade is one of the benefits that can arise from sophisticated exchange systems. Trade is a form of reciprocity. We have seen that food is freely given to relatives and, in most societies, relatives have rights to food. Food is often offered to strangers as a form of hospitality. Though this can be a delicate matter. While food often turns strangers into friends, food not offered on a suitable occasion can turn strangers into enemies[10]. Primitive trading in local markets often involves barter. In situations where people know and trust each other, a mental credit rating often develops. This is common in many rural communities today. I keep a few goats on my land in the Canaries, and if I have too many I give one to the local goat farmer. I do not ask him to pay for it, because I know that one day he will do me a favor in return.

Barter was common in the fifteenth and sixteenth centuries, when the Portuguese, Dutch and English developed trading sea routes to the East Indies. Many of the peoples that they visited to acquire spices did not have a monetary economy, and did not operate a price market. In preparation for the first voyage of the fledgling East India Company "After numerous arguments and deliberations the merchants settled on a mixed cargo of lead, iron, (both wrought and unwrought), Devonshire cotton kerseys, broadcloth and Norwich woolens, as well as several boxes of trinkets ..."[11]. They traded these for provisions on the voyage, and for spices to be shipped home. The latter were paid for by a mixture of money and barter.

An important benefit of trade is that it provides an outlet for surplus production. Agricultural and manufactured produce in one locality can be exchanged for that of another locality. Both parties benefit by widening the range of food and goods that they can enjoy. The downside is that trade usually involves middlemen, who buy at one price and sell at a higher price. In 1621 spices sourced in the Indies had increased more than eightfold in price by the time they reached Aleppo overland.[12] No wonder the maritime nations of Portugal, Spain, England and Holland were anxious to cut out the middlemen. Where I live, in the Canaries, unlabelled wine can be purchased directly from the grower, at the local market, at half the cost of the same wine from a shop, and one quarter of the price in a restaurant.

Quality control

The middleman not only influenced the price of food, but also its quality. As villages and towns grew, the small barter-markets developed into important trading places. As the markets became important for the prosperity of a town, the authorities found it necessary to take precautions against robbery, protection-rackets, or any other threats to the reputation of the market. In the middle ages offences against health and hygiene became a problem, but the authorities were most concerned about quality control. In France langueyeurs examined pigs' tongues for ulcers, and in Venice fishmongers stalls were inspected daily so that stale fish could be destroyed. More difficult to combat was calculated fraud. A common scam in the Middle Ages was selling underweight bread. Bread of a certain price was supposed to have a certain weight. But then, as now, the food inspectors cannot be everywhere, and they are not entirely immune from the temptation to accept 'gifts'.

Adulteration of food became widespread once the middlemen could operate out of sight. For example tea was transported by ships, and stored in warehouses .There was plenty of opportunity to make extra profit by mixing tea with various additives. Fake versions of Chinese green tea were often produced from dried thorn leaves and coloring them with verdigris, which was poisonous. Black Indian tea was bulked out by acquiring used tea leaves from hostels, and coffee houses, and then stiffening them with gum solution and coloring them with black lead. In 1820 Frederick Acum published *A Treatise on Adulteration of Food and Culinary Poisons*. Acum, a German born chemist, was hounded out of Britain by enraged but influential food manufacturers. But the damage had been done, and from 1850 *The Lancet*, regularly published articles describing the extraneous matter that could be found in samples of staple foods, purchased at random from London shops. In 1860 the first British Food and Drugs Act was passed.

By the nineteenth century, the growth of towns, and their supply by road and rail, brought an organized food industry into being. In rural market towns, chickens for the table were sold live. There was no question as to their freshness. When chickens have to be transported a long distance to market, they are likely to arrive in a sorry state, so it is better to transport them as dead birds. Prior to the widespread use of refrigeration, the dead chickens would have to be transported without delay, but sometimes delays were inevitable, so there was always going to be some question about the safety of eating such chicken. Whereas fish could be preserved by salting or drying, a chicken had to be eaten soon after it had been killed[13].

In 1934 John Kimber, a Californian businessman, invented modern chicken farming. Certain breeds of chicken were intensively bred for egg laying, and other breeds were intensively bred for meat. This innovation led to a new sort of food quality problem. By this time regulations concerning hygiene and food adulteration were enforced, and it was no longer profitable take short-cuts, or dilute food products with inedible material. Instead, the emphasis would now be in reducing the costs of food production and distribution. The result would be an increase in factory farming, and a decline in food quality that has persisted to this day. The impact of factory farming of poultry, pork and veal has had three main effects. It has greatly reduced the cost to the consumer. It has greatly reduced the quality of the food that arrives on the average family table.[13]. It has greatly speeded up the evolution of the animals concerned. In other words, our food is evolving faster than ever before[14].

Recently, there has been some attempt to reverse these trends. In the UK farmers markets, where farmers can sell direct to the public, are on the increase, and various food organizations will deliver a weekly selection of vegetables to your door. Where I live in the Canaries, there has been a decline in the amount of land in cultivation over the past forty years. Recently, however, local markets have opened up, and growers can sell their produce directly to the public. The result has been a noticeable increase in the amount of land under cultivation. The growers benefit, because they sell their produce for more than they could previously, and the buyers benefit from lower prices and from fresher food. In other words, the profit-making middleman has been cut out. Unfortunately the scale of this small market revolution is small. Those in the know benefit from it, but the vast majority continues to put up with low quality food. In the next Chapter we find out why this is so.

The changing farmscape

Farming has transformed the countryside in many countries. As we saw in Chapter 4, early Neolithic farmers cleared land for cultivation, using a slash and burn approach, and this practice carries on to this day, in the Brazilian forests, for example. But in addition to the direct physical impact on the environment, the increase in agriculture has had a political impact. Agriculture requires land, and land acquisition leads to land ownership, a notion foreign to nomadic peoples. Early farming families would regard their land as theirs, but families grow into clans, and clans grow into tribes. Jared Diamond describes how the Bantu, living originally (i.e. 5000 years ago) in (what is now) Cameroon and Nigeria, spread eastwards and southwards throughout Africa. Initially they domesticated cattle and yams, but as they moved eastwards into drier areas they took up the local millet and sorghum. In 3000 years they had reached the east African coast, and had acquired iron. Within the next 2000 years they swept all before them, reaching Fish River 500 miles east of Cape Town. In 1702 they clashed with white European settlers who were spreading out from the Cape. There followed 175 years of warfare between the two groups.

The Bantu displaced the native hunter-gatherer Khoisan peoples (some of whom had acquired sheep and cattle) in those areas where they could cultivate the land. It is probable that the Bantu, with their agriculture and their consequent higher reproductive rate (see Cheaper 4) simply out-competed the native population. There would be some trading and inter-marriage between the two peoples, each adapted to different habitats. However, the important point is that the habitat occupied by farmers becomes territory. Farming is static, and farmers come to think of themselves as owning the territory.

Equivalent scenarios have taken place in many parts of the world, at very different times[15], but the progression is fundamentally the same: Nomadic groups are displaced by settled communities: Territory becomes established and inevitably disputed: Territorial defense requires a new kind of social organization, involving division of labor, specialization and professionalism not directly associated with food production: Political control by the few of the many, includes political control over food production and distribution.

Once land comes under political control, it is changed forever. It is no accident that many of the roads in Britain, and other countries, date back hundreds of years. For example, the Ridgeway has been called the world's oldest road[16] .Some towns and villages have disappeared but many, throughout the world, have existed for thousands of years. Archaeologists find traces of ancient settlements on land that has remained largely unchanged, but most of the land in some countries, and much of the land in others, has been changed forever by the activities associated with agriculture, including fields, roads, settlements and markets.

Environmentalist and campaigner Colin Tudge[17] has identified many of the factors contributing to our present problems concerning land and food supply. One of these is farming for money. Imagine that it was possible to grow money on trees. What would farmers grow? Would they grow money or would they grow food? In Afghanistan, many farmers grow money, in the form of opium poppies. In other parts of the world, they grow money in the form of tobacco. If everybody grew inedible crops there would be a food shortage. In fact in many parts of the world there is a food shortage. As Colin Tudge points out "Agriculture is just a business like any other." As such it is subject to business pressures, including corporate buying power and globalization. Consequently, for example, many British farmers cannot make a living, because so much cheap food is imported. The problem for the buyer is that, while much of the food in cheap price-wise, it is also cheap in quality. It is very much a case of buyer beware. Inexpensive food may make you feel prosperous in the short run, but eating poor quality food will make you feel poorly in the long run.

So why to people do it? It is clear that people do, because you only have to walk down the street to see the growing malnourishment of the people there (i.e. obesity, drawn faces, very red faces, etc). Many people buy their food in a supermarket because it is less expensive in the supermarket than elsewhere. When they pay a low price for a chicken they think it is a fair exchange. They are deceived, because it is not a simple exchange of chicken for money, as it would have been years ago. The buyer in the supermarket is indirectly paying for the food transport, and the food storage and inspection. They are paying in terms of the cost of fuel (i.e. the cost of fuel in traveling from their own home), and in terms of the carbon footprint of the food. So the real exchange equation is like this: one chicken = money cost of chicken + malnutrition cost + carbon footprint + cost of fuel for home and car. So why do people do it? Part of the answer lies in advertising, and to this we turn in the next Chapter.

The trader will say that, to compete with other traders, they have to provide what the customer wants, and the customer wants cheap chicken. Maybe so, but how does the customer know what they are buying? The customer has to trust the trader, and the trader is under an obligation (these days) to tell the truth. But the traders do not always know the whole truth, and even if they did, would they – could they – tell the whole truth?

This is an age-old problem, going back to the origins of trading. So here you are. The Neolithic revolution has started in your part of the world. You live a settled agricultural existence, and your diet is high in carbohydrate. You live in an area where there is not much meat available, so you get some of your protein from beans. You also cultivate fresh-water fish, in artificial ponds. So mosquitoes breed where you live, and your locality has a reputation for certain types of disease and sickness. However, you and your kin seem to be untroubled by such problems, and this means that you live in peace, free from invasion by foreigners wary of malaria. One day a trader comes to your locality with a new kind of bean. They are said to be good and would be easy to grow. You all agree that you should plant and harvest these beans, which is what you do. The beans are a great success, they are delicious. After a short while some of the men in your community, including you, start to feel unwell. You become listless and pale, you have breathing difficulties and you die (from favism[18]). This is our fifth case of *death by eating*. The beans (called broad beans or fava beans) have an effect upon some genetically susceptible people, who have a particular enzyme deficiency. This deficiency (called *favism*) confers protection against malaria. In other words, you were able to live in your locality because you had a genetic protection against malaria. The price you paid for this protection was that a certain food (fava beans) was poisonous. This did not matter so long as these plants were not available. Once they became available, as a result of trade, you were in danger. Blame the trader. Did the trader know, and if not, why not? Very probably, the trader did not know. In some cases, of course, the traders do know that their commodities (e.g. cigarettes or drugs) are dangerous.

Chapter 6   *The power of advertising*

In nature, advertising is a form of display, evolved by plants and animals as a means of communication. As we saw in Chapter 2, plants may display brightly colored, or distinctively scented, flowers to attract pollinating insects. Animals may display warning coloration, and warning noises, to deter predators. Sexual display, often shown by males holding a territory, serves both to attract females and ward off rival males. To be effective, such displays should be conspicuous and oriented towards likely recipients. The advertising type of bird song, for example, should be easily locatable, and should indicate the species, sex and motivation of the singer. Thus advertisement by plants and animals can take many forms, each designed by nature to be maximally effective.

Many animal displays are derived from simple behavior patterns through an evolutionary process called ***ritualization***, that makes them more stereotyped, conspicuous, and reliable as a means of communication. Some have evolved from intention movements and conflict postures. Others have evolved from defense behavior, and from the physiological responses involved in mild stress. Examples in humans include blushing and smiling. Both are culturally universal displays that are stereotyped, both in their detail and in the situations in which they occur. Smiling is made more effective by the distinctive coloration of the lips. In some cultures this is enhanced artificially. In this chapter you are a modern city-dweller. You work an eight-hour day, and a five-day week. You are under a little stress at work, but who is not? You are not one of those people who are susceptible to advertising – are you?

Advertising as a means of communication

Most advertising takes the form of *broadcast* communication. But what is communication? In biological terms, communication is the transmission of information from one individual to another, which is designed to influence the behavior of the recipient. This means that natural selection must have acted on the sender to fashion the signal, and of the recipient to detect the signal. Thus fortuitous information transfer does not count as communication. For example, when a cow lifts its tail prior to defecation, and the animal behind moves out of the way, we would not say that the cow is communicating with the animal, because tail-lifting functions to promote cleanliness, and has not been designed as a signal to others.

In a simple communication system, a *source* encodes and transmits a signal, which is detected by a receiver, and decoded into meaningful terms. Encoding involves transformation of a message from one representation to another by operation of code rules. Thus an angry dog may bare its teeth. This *signal* is the physical embodiment of a message, or set of *signs*, by which an animal influences the state of another animal.

In the natural environment, it is important to distinguish between *transmitted* information and *broadcast* information. Transmitted information is measured in terms of an increase in the predictability of the receiver's behavior following activity by the source, or sender. In the case of broadcast information, the source emits an untargeted signal, by means of which an observer gains information about the source. For example, there are three cockerels on my land - a white one, a black one and a bantam. I can distinguish one from another by their pattern of crowing. The crowing is not targeted at me, but I still gain information from it. The crowing of a cockerel is a kind of advertisement about the identity of the source. In other words, the crowing cockerel is broadcasting information (this is me) to the world at large. Thus broadcast information is a measure of the information obtained from a signal by an observer.

Most advertising is a form of broadcast information, but most advertisers would dearly love to employ transmitted information – information that is targeted at potential customers. Their problem is that the effectiveness of transmitted information is measured in terms of an increase in the predictability of the receiver's behavior. It is difficult for advertisers to gauge how effective their signals are, but as we shall see, they have their ways and means.

Historically, advertising goes back hundreds of years. Commercial messages and political campaign displays have been found in the ruins of ancient Arabia. Wall or rock painting for commercial advertising is another manifestation of ancient advertising, and it is present to this day in many parts of Asia, Africa, and South America. The ancient Egyptians used papyrus to create sales messages and wall posters, and this was also was common in ancient Greece and ancient Rome.

As printing developed in the 15th and 16th century, advertising appeared in the form of handbills, and in the 17th century advertisements started to appear in weekly newspapers in England. These early print advertisements were used mainly to promote books and newspapers and medicines. The father of modern advertising is generally agreed to be Thomas Barrett, who married into a famous soap making family and realized that they needed to be more aggressive about pushing their products if they were to survive. He launched the series of visual advertisements depicting cherubic children, images taken from 'fine art' and used them to connote his brand's quality and purity. The result was the famous Pears Soap advert.

In the natural world, the effectiveness of communication is influenced by the physical environment, by the nature of the receiver, and by the influence of other signalers. Different sensory modalities are best suited to different habitats, as human advertisers have realized. For example, in an open area visual communication can be effective, but in a woodland vocal communication is better. On a noisy underground station, or on the train, the ads are visual, while in the quiet of your home they are also auditory, coming over the radio, or from the television.

The nature of the receiver is also important, both in the natural world and in the human situation. A message is effective only in so far as it is tailored to the sensory apparatus and receiver-psychology of the recipient. The receiver psychology embraces the detectability, discriminability and memorability of the signal. So an advert must stand out from the background. It must be easily discriminated from other messages, and it must be memorable. Moreover, the effectiveness of communication is influenced by other signalers, particularly where there are closely related. Advertisements for cars tend to be all of a kind, so that the receiver of the information easily confuses different makes of car. The same goes for washing powder, and cigarettes. The same principle holds in nature. Song birds are all saying the same thing – this is my territory – but the songs of different species are distinctive. Even humans can tell the difference. It would seem that human advertisers have something to learn here.

Advertising as manipulation

The power of advertising lies in getting others to do what they might not otherwise do. Advertising has a number of aspects that we should look at. The first is providing information. An

93

animal, or person, the putative customers, will not do what they might not otherwise do unless they have information about the alternatives. The prime function, and the historically early function of advertising, is simply to provide information about the availability of the product. So a flower opens when it is ready to provide nectar, and closes at other times. Similarly a shop opens for business at certain times, and closes at other times. The flower and the shop advertise the fact that they are open, i.e. they provide the information to the customer.

Of course, the customer may ignore the information and do what they were going to do anyway, so the next function of advertising is to persuade. To persuade an insect to visit, a flower must be attractive to that particular type of insect. As we saw in Chapter 2, insects tend to be fussy about the flowers that they visit. They budget their time and energy, and do not waste time and effort on any old flower. To persuade a passer-by to enter the shop, the shop must be attractive, not to all-comers, but to those likely to spend money.

Thirdly, there is *manipulation*. In the biological sciences manipulation occurs when a receiver obtains information about the signaler, against the interests of the signaler. For example, the courtship display of a male may attract rivals. The rival may benefit by receiving the signalers message to the female, because it leads him to the female, against the interests of the signaler. This form of manipulation has been dubbed 'eavesdropping'.

Another form of manipulation is *deceit* – the signaler's benefit increases at the expense of the receiver. For example, hoverflies have no sting, yet they have the black and yellow warning coloration of a wasp. Predatory birds that avoid the wasp because of its sting, also avoid hoverflies. So the signaler, the hoverfly, benefits at the expense of the receiver, the bird, without carrying the expense of developing a sting. However, if hoverflies become more common that wasps, then some birds will not learn to avoid them (because they have not learned from a wasp encounter), and some birds will devalue the warning information, while others may learn to discriminate between wasps and hoverflies. As we shall see, the same principles apply to human advertising.

Finally, there is manipulation by brainwashing. This too occurs in nature. The parasitic protozoan *Toxoplasma gondii* reproduces in a cat's stomach. To get there it infects a rat and enters its brain. There it manipulates the rats' response to the smell of cat, making it attractive rather than to-be-avoided. Such rats are more likely to be eaten by cats, so the parasite benefits as a result of manipulating the rat. Then there is the lancet fluke (*Dicrocelium dendriticum*), which reproduces in the stomach of a sheep or cow. It does this by infecting an ant, and manipulating its brain so that the ant has a persistent urge to climb up blades of grass in the evening, and stay there throughout the night, rejoining the colony the next morning. This behavior makes the ant more likely to be eaten by a cow thus helping the fluke to complete its life-cycle. There are numerous other examples of direct behavior-manipulation by parasites.

Just because an advert is persuasive, does that mean that it is manipulative? It depends. If an advert is deceitful, then clearly it is manipulative in the sense that the message deceives the buyer into buying something different than they thought they were buying. When a bird lays an egg in its own nest, they 'expect' that they are buying one of their own offspring. But then along comes a cuckoo that turfs the egg out of the nest and replaces it with an 'identical' one of its own. The parents are in fact buying an offspring of another species. The cuckoo has deceived them. Initially, the advert is the egg. It looks genuine. Then comes the real ad. The hatchling is a lovely big one, and it has an irresistible, supernormal, gape. Who can resist dropping hard-earned food into that mouth?

Luckily for us, these days, deceitful ads are illegal, and there is a watchdog on the lookout. In the UK there is the Advertising Standards Authority. This is a self-regulatory organization, funded by the industry. So far, it remains independent, with the intent of upholding certain standards or codes. However, if the government becomes dissatisfied with the performance of the ASA, it can always threaten, or enact, appropriate legislation. Similar arrangements exist in many countries – such as New Zealand, South Africa, Canada, and many European countries. Where the advertising industry operates a system of self-regulation. advertisers, advertising agencies and the media agree on a code of advertising standards. These self-regulatory watchdogs do not have many teeth. In the UK, the ASA cannot fine companies who breach its code, or bring legal action against bad advertisers. Sanctions such as bad publicity are available, and can be effective. In the last resort the ASA can refer a case to the Office of Fair Trading. This is the ASA's legal enforcement body. The OFT has powers to fine companies and take legal action against them. If the ASA has trouble with a repeat offender, they can refer the matter to the OFT under the Control of Misleading Advertisements Regulations 1988. The OFT can then take whatever action they deem necessary. In the USA there have been increasing efforts to protect the public interest by regulating the content and the influence of advertising. In particular, there is a vigorous debate on whether, and how, advertising to children should be regulated. The evidence suggests that food advertising targeting children is an important factor in the epidemic of childhood obesity in the United States of America.

Advertising addictive substances amounts to manipulation. These days, the term *addiction* is used to describe a recurring compulsion by an individual to engage in some specific activity, despite harmful consequences to the individual's health, mental state or social life. Just like the ant that climbs the blade of grass, or the fearless rat (above), the individual is compelled to behave in a certain way, whether it be drug-taking, gambling, smoking or overeating. Advertising that entices people to become addicted is a form of manipulation, and governments are becoming more and more aware of the dangers involved. Possession of certain substances, trading in such substances, and advertising their availability, is illegal in many countries. In other words, the authorities are aware that these substances are addictive and damaging, and they attempt to prevent their citizens from being manipulated. Unfortunately, prohibition often has undesirable consequences. Prohibition of alcohol the USA (1920-1933) was undertaken to reduce crime and corruption, solve social problems, reduce the tax burden created by prisons and poorhouses, and improve health and hygiene in America. Although consumption of alcohol fell initially, it subsequently increased. Crime increased and became organized. The court and prison systems were stretched to the breaking point; and corruption of public officials was rampant. Prohibition removed a significant source of tax revenue and greatly increased government spending. It led many drinkers to switch to other dangerous substances. It also had deleterious effects on business, for example it destroyed what was a fledgling wine industry in the United States, an industry that took forty years to recover. Prohibition of smoking in public places has been imposed in many countries in recent years, and consumption of cigarettes has fallen in those countries. Whether or not this will be maintained remains to be seen, but the fact that smoking is seen to be deleterious for non-smokers, may be an important factor is molding public opinion.

The effects of food advertising

The idea behind food advertising is to increase demand for the particular items being advertised. **Demand** is the quantity of a good or service that people want to buy. The extent of demand depends upon a number of factors[1], but other things being equal, when demand for one type of food goes up then demand for another must go down, provided the population is static. There are limits to what people can eat and store, so if a certain type of food becomes all the rage, then some other foods must decline in popularity. So advertising aims to promote one type of food at the expense of another.

If the population is increasing, then the demand for food will increase. If supply cannot keep up with demand, then the price of food will increase. As we shall see, the world population is increasing, and world food prices are increasing, and this suggests that there are problems with the supply of food.

In country markets, such as the one in the village where I live in the Canaries, food is advertised by being put on display, together with the prices. You can handle the fruit and vegetables, you can taste the cheese and the wine. In the supermarket in the nearby town, the food is also put on display, but you cannot touch it, because it is pre-packaged. You cannot taste it, and I have to put one my spectacles to read the price. In doing so I discover all sorts of things about the food on sale that I would rather not know. I can discover where it has come from (often a long long way). I can discover what has been done to it (e.g. radiation, preservatives and various additives). I can discover all sorts of reasons why I would not want to buy it.

In Shepherds bush market in London you can buy a scraggy-looking boiling fowl for £1. In a supermarket in the same street you can buy a plumped-up chicken for £2-3. Weight for weight they probably cost the same, but one tastes so much better than the other. Guess which one – the one that had some exercise when it was alive; the one that had some variety in its food; the one that experienced a normal day-night cycle; the one that had a longer and better life. Guess which type of chicken is more popular. For which is there more demand? Not the tasty happy chicken, but the plump sad one. How can this be?

The answer lies in advertising, the persuasive power of advertising. But we don't see numerous ads for chickens on TV, and we don't hear about them on the radio every day. No, but we do get plenty of supermarket ads. What is advertised is the shopping place. A supermarket is super – super clean, super cheap, super convenient. When you enter a supermarket, everything looks lovely. Everything is lovingly presented. They even have music. The chickens are all lined up on display. Ordinary chickens for £3, free-range for £5, and organic ones for £8. These are the show chickens. What you don't see are the reject chickens and reconstituted chicken. You don't see these, because they hidden in the ready-meals (just warm it up), the sausages, pies and the pet food.

We all know this. We all know that food quality is gradually on the decline. We are bombarded with this information by TV Chefs, magazines, and the health food and organic freaks. So why do we buy it? Why do we buy the advertising message, and why do we buy poor quality food?

Part of the answer lies in our food psychology, as we see in the next Chapter, but the conventional answer is that supermarkets are cheaper and more convenient. Whether it is true that food is cheaper in the supermarket is open to question. As Colin Tudge remarks

"I shop when I can at a couple of traditional markets where beans are sold en masse in paper sacks and vegetables come straight out of boxes and it all costs about a third, and sometimes only a tenth, of what the same things would cost in supermarkets. Food is as dear as it is largely because of middle-men's mark-ups. Take away the fripperies, and the cost can come down dramatically."[2]

Colin Tudge lives in Oxford, so he is presumably talking about the Oxford Supermarkets which are mostly on the ring road. If you try to buy food in central Oxford, it is more expensive than in the supermarkets. Moreover, food from the supermarket is usually less expensive than it is in the corner shop, many of which are unfortunately closing down. Supermarkets often advertise on the basis that their food is less expensive, but their comparisons generally relate to other supermarkets.

Supermarkets are more convenient because you can do all your shopping in one place. If you visit a supermarket once a week, and pile the week's food into your car, it is probably true that you spend less time shopping (per week) that you would otherwise. But what if you have no car? What if you are too old, too disabled, or too poor to have a car? Then the out-of-town supermarket is not so convenient. Moreover, the handy corner shops are going out of business. So supermarkets are convenient for some people, especially people who are short of time. At least, this is what they say, but is it true?

If an urban office worker works an eight-hour day, and sleeps an eight-hour night, there are still eight hours left in the day. That leaves 72 hours of leisure per week (weekends included). The hunter-gatherers that have been studied (see Chapter 2) also work a forty-hour week, so they also have 72 hours of leisure, of which they spend 20 hours per week preparing food and cooking. In other words they spend less than three hours per day preparing food, cooking and eating. If the urban office worker chose to spend 2 or 3 hours per day preparing food, cooking and eating, there would still be plenty of time for other leisure activities. But the urban office worker claims that there is not enough time in the day to 'waste' on food preparation and cooking. There are better things to do, but at what cost to the body?

Let us look at just one of the common ingredients of fast food, and of supermarket ready-to-eat food (just warm it up), namely chicken nuggets. In 2006, in the UK alone, about 120 million chicken nugget meals were eaten at home or as take-away food. This figure does not include the millions served up to children at school, and patients in hospitals. So what do chicken nuggets contain? Chicken meat is, of course, a prime ingredient. The food label usually gives the proportion of chicken meat, but what counts as chicken meat? The answer is chicken skin + reconstituted meat slurry[3] + mechanically separated meat[3] + processing additives + one harmless ingredient, which is water, sometimes making up 40% of the total. Analysis of chicken nuggets has revealed that many contain a high proportion of chicken skin, undeclared bovine proteins, soya proteins, gums, illegal antibiotics, phosphates, and maybe some chicken breast meat[3]

Some of these ingredients are not very good for you, but provided the nuggets are prepared under hygienic conditions, they will not cause a quick death. What happens to your body in the long-term is another matter.

There are urban people in Britain who do not eat chicken nuggets. It can be done. The Clarks (successful cookery writers and restaurateurs) rented an allotment in the East End of London, near the Grand Union Canal. They discovered the other allotment holders, including immigrant Turks and Cypriots, grew and cooked an extraordinary range of ingredients. These people spent time in their allotments. They spent time cooking and eating. Their food is wholesome, delicious, and fresh[4].

So what about you? You are not susceptible to advertising. You are sensible about your use of time and money. Nevertheless, you are sometimes in a hurry, and you sometimes buy pre-packaged food without really thinking about it. You sometimes buy packaged food without reading the label, and you sometimes buy fast-food and eat it on-the-hoof.

Besides, your garden is not big enough for you to grow your own vegetables, so you buy them from a supermarket. Your refrigerator is too small to stockpile food from a prolonged shopping trip, so you tend to buy just enough for one or two days. It is all very rational, and suits your lifestyle. There is nothing to worry about, so you don't worry. You are a little overweight, but not much different from everybody else. There is safety in numbers. Then you have a stroke – not to worry – about 150,000 people in the UK have a stroke every year.[5] Why did you have a stroke? There are many causes of stroke, but you had one because you are overweight, have high blood pressure, and have too much salt and too much fat in your diet. Also you do not eat enough fresh fruit and vegetables. But how were you to know - these things are not advertised? They are advertised. They are advertised in almost every doctor's surgery and every pharmacy. It is true that they are not advertised as loudly, or as effectively, as supermarket and fast-food ads. But then, you are not affected by these, are you? Anyway, not to worry – lots of people have strokes – safety in numbers. You had your first stroke ten days ago, and now you have your second. Now you are dead[5] This is our sixth case of *death by eating*.

Chapter 7   Food and the family

"Eating patterns emerge in childhood and may set the stage for the healthfulness of eating patterns as adults. Parents are important in shaping young children's eating patterns because they select foods of the family diet, serve as an example through their own behavior and provide direct instruction about when, what, and how much to eat. Studies over the past two decades have shown that children acquire preferences for foods to which they are routinely exposed. As a result, children choose to eat foods that they are served most often, and prefer what is available and acceptable in the home. Because eating habits are difficult to modify once established, it is important that parents and practitioners understand how healthful patterns of eating can be encouraged. Establishing a pattern of good choices in childhood is likely to carry over into healthier habits in later life."[1]

There you are, straight from the pediatricians' pen. It is simple, give your children what you want them to eat, and they will be influenced for the rest of their lives. If they end up with bad eating habits, you only have yourselves to blame.

There is obviously something in this. You only have to walk down Shepard's Bush market in London to see the huge variety of vegetables from all over the world – much greater than the variety that you see in a supermarket. Why is this? – Because the people who visit the market originated in (i.e. their parents or grandparents came from) Africa, China, India, the West Indies, and all sorts of other places. These are not first-generation immigrants. They are second, or third, generation. Yet they still prefer to eat their ancestral food. Can this can only be due to parental influence?

As a biologist, I can confidently say that no behavior is solely the product of parental influence, or solely the result of any one influence. In animals in general, there are genetic influences, food imprinting, conditioned aversion, social facilitation, and politics. Translated into human terms this roughly amounts to genetic influences, parental influence, peer pressure, and status. Sure enough, we find all these influences promulgated by scholars and scientists of various sorts. The situation is complicated, so let us take these factors one at a time.

Genetics and food preferences

You are born with your genetic makeup and there is not much you can do about it. Genetic variation in human taste has been known for a long time[2]. To some people coriander tastes 'soapy', while other people dislike caraway, but they do like the closely related cumin, fennel, etc. These reactions are due to particular taster genes. Some people have heightened taste sensitivity, and they are sometimes called supertasters[2]. On the whole, super-tasters are *less* likely to enjoy, and thus consume, certain foods, than are ordinarily people. Documented examples for such lessened preference and consumption include alcoholic drinks, lettuce, coffee, grapefruit, green tea, spinach, and soy products[2]. So it is fairly clear that genetic variability in taste can affect people's food preferences.

In experiments with school children, I discovered that children (age range 13-16) asked to rate small sandwiches, on a like-dislike scale, fell into two distinct groups: those who liked both parsley and coriander sandwiches, and those who liked parsley but disliked coriander. In a similar experiment, there were also two distinct groups: those who liked bread containing cumin seeds and bread containing caraway seeds, and those who liked

104

bread containing cumin seeds, but disliked bread containing caraway seeds. The proportions of likers and dislikers were (in both experiments) consistent with the hypothesis that there was a single gene responsible for the antipathy towards coriander and caraway[3].

Some children may be extra sensitive to bitter tastes. Genetic studies involving a bitter gene (those with the gene are more sensitive to bitter things) show that children who carried the bitter gene were much more sensitive to the bitter taste than their mothers, even though their mothers also had the gene. So childhood may represent a period of heightened bitter taste sensitivity (in some children) that lessens with age. Children who have the bitter taste gene tend to prefer higher levels of the sugar in their diet, and they may become conditioned to sweet-tasting food (see below). If your child dislikes spinach, it may be that they are super-sensitive to bitter tastes.

In Chapter 4 we saw that humans vary in their ability to digest milk as adults, and the lactose intolerance involved has a genetic basis. Clearly, this genetic effect is likely to influence their food preferences. If you are lactose intolerant, and you drink some milk, you will have unpleasant and maybe severe digestive consequences. You will quickly learn not to drink milk again. In Chapter 1 we saw that animals that are unable to detect essential vitamins and minerals either by taste or by their levels in the blood can develop strong preferences for foods containing the missing substances. They are able to do this as a result of their behavior towards novel foods. Many animals exhibit *neophobia*, a wariness of novel food. Often they will take a small sample, and wait a while before taking more. If the results from sampling novel food are bad, you avoid it in future, if good, you incorporate it into your diet. This type of learning can be very rapid. Studies of children show that *neophobia* is highly heritable (has a strong genetic component)[4]. In other words, there is a genetically controlled tendency to be wary of novel food, and this is likely to vary from child to child. So, all in all, a child may be fussy about food as a result of an inherited taste disposition, an inherited digestive intolerance, or an inherited wariness of strange food.

Food imprinting

Imprinting is an aspect of learning that takes place during a particular sensitive period in the early stages of an animal's life. For example, lambs follow the person that has reared them on a bottle. Even after the lamb has been weaned and joined the flock, it will approach its former keeper, and try to stay near by. If the lamb has grown into a mature male, it may show a sexual interest in its keeper. The lamb is imprinted upon the keeper, and this has both short term and long term aspects. The lamb follows the keeper when young, and as an adult is shows some attachment to its keeper.

The following response is shown by many precocial species, which can run around soon after birth. The juvenile initially shows a fairly indiscriminate attachment to moving objects. Thus ducklings separated from their mother will follow a crude model duck, a slowly walking person, or even a cardboard box that is moved slowly away. This type of attachment takes place during a sensitive period that varies from species to species. Ducklings most readily form an attachment to a moving object between 10 and 15 days of hatching, but the sensitive period lasts for another six weeks. Imprinting will not occur if first exposure occurs after the sensitive period. In addition to its effects on parent-offspring relationships, imprinting in animals can have marked effects upon the social relationships as adults, and on food selection and habitat selection. Do the same principles apply to food? Food imprinting in animals has been shown to occur in a few species.[5]

Whether true food imprinting occurs in humans is not entirely clear. To count as true imprinting, there would have to be a sensitive period during which food preferences were established through learning, and these preferences would be long-lasting. Food preferences established after this period would not count as imprinting, and here lies a problem. Humans establish food likes and dislikes throughout their lives, and the extent to which early food preferences persist is unclear. There is some epidemiological evidence suggesting that the quality and quantity of nutrients available during prenatal and early postnatal development can affect susceptibility to various adult-onset chronic diseases, such as cardiovascular disease and type II diabetes. Birth-weight, in particular is related to these factors, and such studies have led to a *fetal origins* hypothesis, the idea that the mother's nutrient intake can affect the child's lifetime dietary preferences and susceptibility to certain diseases. However, it has proved difficult to separate the effects of early nutrition from genetic factors[6], and although some workers claim that food imprinting occurs in humans as well as other animals, there is no good evidence that the infants develop their early food preferences through learning, which would be necessary for there to be true imprinting.

Children of obese parents have a substantially higher risk of adult obesity than children of lean parents. Adoption and twin studies have shown that this risk is largely genetic. Comparisons of energy intake or expenditure in children of obese and lean parents have produced mixed, but generally negative results. There were small differences in food preferences, both in taste tests and preference records. High-risk children ranked the higher-fat foods more highly and had lower liking for vegetables, but there were no differences in the frequency of intake of high-fat, low-fat, high-fiber or low-fiber foods. The children of obese families were rated as enjoying sedentary activities more and were less active than other children. They spent more hours at the computer and watching TV.

Twin studies[6] indicate that genetic factors affect many aspects of eating such as the number, timing and composition of meals as well as degree of hunger and sense of fullness after eating. So it seems that much of the similarity between parents and their offspring, in eating habits, obesity, and susceptibility to particular diseases, is due largely to genetic factors, rather than to an imprinting-like phenomenon. This is not to say that traumatic experiences such as starvation, disease, or parental smoking or alcohol consumption have no effect. They clearly do, but these are not factors that are relevant to our enquiry about normal food preferences and eating habits.

Coonditioning

Although food imprinting proper may not occur in humans, there is another form of learning that is probably responsible for the eating habits that we acquire in early life. Conditioning is a form of learning, in which an animal forms an association between a previously significant stimulus and a previously neutral stimulus or response. In classical conditioning an association may be formed between a significant stimulus, such as the sight of food, and a neutral stimulus, such as a flashing light. Initially, the animal responds only to the food, (e.g. by salivating). If the food is presented together with the light on a number of occasions, then the animal comes to associate the light with the food, and will eventually salivate even if the light is presented alone. The response of the animal is said to have become conditioned to the light.

Now when you eat your food, you do so in a particular situation, a situation that is characterized by various stimuli. By eating in the presence of these stimuli, especially if you repeatedly do so, you come to associate the stimuli with the rewarding consequences of eating.[7] These stimuli include the sight and smell of the food that you are eating. Of course, if the consequences of eating a particular food are unpleasant, then the conditioning will have a negative effect (called food-aversion conditioning, see Chapter 1). Normally, however, the consequences of eating are pleasant and rewarding, and positive conditioning occurs. So while animals have evolved rapid negative conditioning to the taste of food protection against poisons. (as we saw in Chapter 1), there is a reverse side of this coin. Certain tastes readily induce rapid positive conditioning in humans, and the most important of these are the sweet and salt tastes. In nature, salty food items and sweet food items are rare. It is advantageous for an animal to rapidly learn which items are sweet and which salty, so that it can readily take the opportunity to grab such items when the occasion arises. In fact[8], there is evidence that many animals have an open-ended preference for sweet things. The sweet taste signals sugars, and these provide quick energy. For many, it is a case of grab-it-while-you-can when such rare items are discovered. For hunter-gatherers, the sweet foods would include honey and certain fruits when in season. From an evolutionary point of view, an open-ended preference for sweet things would be a good strategy for situations where sweet foods are in short supply. In other words, sweetness is a supernormal stimulus, a stimulus that surpasses a natural stimulus in its effectiveness. . Animals often have an open-ended preference for certain stimuli, such as size, sweetness, or color that is limited in the natural world. Thus herring gulls prefer to retrieve larger than normal (dummy) eggs into their nest. When presented with an artificial supernormal egg, they prefer it to the natural one. To them, a bigger egg is better, and likely to be more viable. For us today, sweet foods are not in short supply. The pioneer nutritionist, John Yudkin[8,] drew attention to the view that there have been two revolutions in the evolution of food. The first was the Neolithic revolution and the second was the Industrial revolution.

"The first was the development of methods of food production, storage, preservation and distribution. This made food much more available to the industrialized countries with their increasing wealth, and made it available too entirely independent of geography and of season. As a result, for the first time in the story of man or indeed any other species of animal, there is now a small but significant proportion of the world's human population that can look forward to living their whole lives without ever really being hungry. In this way, because of the availability of more food, and especially of more and more nutritious foods like meat and fruit, the industrial revolution could have been expected to have produced nothing but an improvement in man's nutritional health."[8]

But it did not. It did not because food production became just another business, involving anything that would make money, regardless of the consequences. In the forty years since John Yudkin wrote the above words, he campaigned against the over-selling, and over consumption, of sugar; and he was targeted by the sugar industry, just as Frederick Acum, 150 years before him, was targeted by the food manufacturers (see Chapter 5). In the forty years since John Yudkin wrote the above words, there have been various attempts by the government to make sure that the food-chain is safe, culminating (in the UK) in the Food Standards Agency, set up in 2000. The very fact that these measures were thought necessary indicates the extent of the influence that the food industry has on our eating habits.

The influence of the food industry is due partly to advertising (see Chapter 6), but largely to 'providing what the customer wants.' Just as a flower provides what an insect wants, in order to induce it to transport pollen, so the food moguls provide what people want in order to condition them to develop particular eating habits. Needless to say, the eating habits of the general public are very profitable for the food industry. Many animals are easily conditioned to prefer particular foods, and this makes evolutionary sense, because once a food has been consumed a few times without deleterious consequences, then the animals learns that this food is safe. Humans are no exception. We have natural defenses against food poisoning, indigestion, etc. There include our taste mechanisms, our wariness of new food (neophobia), and our rapid food-aversion conditioning. But these defenses against the short-term affects of unsafe food do not protect us against the long-term effects of our eating habits. Food manufacturers know that they will be in big trouble is they provide unsafe food, and they make sure that their standards of hygiene, food storage, etc. are above criticism. There are occasional lapses, but these are minor matters, when you consider the scale of the food business. But food manufacturers also know that people are very habit-prone when it comes to food, and that there is money to be made by providing the type of food people want as cheaply, and profitably, as possible.

In 1969 John Yudkin observed that,

"The industrial revolution ... has made available to a substantial number of people, on the one hand a better range of nutritionally desirable foods than has ever before been possible, and on the other hand nutritionally undesirable foods that previously were consumed in much smaller quantities. For these people the result has been simultaneously the almost complete elimination of nutritional deficiency disease, and the production in virtually epidemic form of a range of diseases

111

that were previously quite rare. "The malnutrition of affluence" is more than a striking catch-phrase."[8]

So what is involved in the malnutrition of affluence? Everybody has their favorite foods, for breakfast, for lunch, for dinner, or whatever. So why is your favorite your favorite? The answer is that, there are many factors that contribute to your preferences and your dislikes: your genetic inheritance, your upbringing, and simple conditioning. You eat a certain food in certain surroundings, and with certain consequences, and this leads to an association between these various stimuli. Many people find that their cat tends to conservative about its food. A cat that is fed a certain brand of dry food, day after day, in the same place, will often reject equivalent food of another brand. It has become conditioned to that particular food. Eating a particular food, in a particular place, at a particular time, has become a habit. Many of us are prone to such eating habits, just like our cats. We are easily lulled into a sense of false-security.

If you want to wean your cat off its expensive favorite food, then do it very slowly. Feed it in the same place each day, and at the same time each day, and then gradually add some alternative food into your cat's diet. The cat is conditioned to its favorite brand, but such conditioning is very malleable. Similarly, if you think that you yourself are taking too much sugar in your tea, or putting too much salt on your food, then cut down very slowly. After a few weeks you will find yourself taking less sugar or salt and not noticing the difference. In fact you will find that food that used not to seem sweet now seems too sweet, and foods that used not to seem salty now do seem too salty. It is this aspect of conditioning that is exploited by food manufacturers. By adding unnecessary amounts of salt or sugar to their products they condition their customers to prefer their products over others.

Additives

Food additives are substances added to food as preservatives or to improve taste or appearance. Some additives have been used for centuries. Historically, food has been preserved with salt (see Chapter 1), vinegar, smoke, and nitrites. Unlike food adulteration (see Chapter 5), food additives were applied with benign intent; but that is not so true these days, as we shall see. Benign food additives include seasoning (pepper and salt), culinary herbs and spices, sugar, etc. The idea behind such additives is to improve the taste and smell of the food. The flavor of food is a combination of its taste and odor (see Chapter 1).

Nowadays there are additives designed to alter the appearance of food. So pastries are often sprinkled with very fine sugar, and pies may be given an egg-glaze, etc. Sometimes, however, the appearance of food is altered or preserved with additives that are less than desirable. For example, when lettuce leaves are cut they quickly turn brown. This discoloration can be prevented by spraying the lettuce with sulphites, and this additive is used by restaurants and supermarkets that sell packets of ready-to-eat-lettuce.[9]. Butter varies in color, but it is invariably dyed yellow. Red meat that has been cut quickly goes brown. Some stores use carbon monoxide to delay such discoloration. Thus the meat looks fresher than it really is.

Additives are also used to alter the texture of food. Thus lecithin is added to margarine to make it more spreadable, and grains of table salt are coated with chemicals to prevent them sticking together. Ice cream has so many additives that is has become unrecognizable.

"One typical brand of ice cream at present on the market is composed of milk fat, corn syrup, cheese whey, monoglycerides and diglycerides, carob bean gum, cellulose gum, polysorbate 80, carrageenan, natural and artificial flavor, and artificial colour."[9]

In our house we make ice cream without a single one of these ingredients. Indeed, it is not difficult to make ice cream from ingredients taken directly from the garden (provided you have chickens). So why all the chemistry? It all has to do with marketing a product with the cheapest possible ingredients, with an ability to withstand transport and storage, and with a facility for hoodwinking the customer.

We tend to think that the days of harmful additives are over, that government-funded regulatory watchdogs (see Chapter 5) have banished such food adulteration. But recollect that as recently as 1981 Spanish cooking oil, containing industrial rapeseed oil killed 165 people, and in 1985 Austrian wine was sweetened with diethelyne glycol (an ingredient of antifreeze) with fatal consequences. Nowadays the food and drink manufacturers, like urban foxes, are much more wary and much more cunning. Moreover, governments are finding it increasingly difficult to reign in the international food-producing corporates. When tomatoes 'packed in Italy' are actually grown in China (think pesticides), and half of the 'Basmati rice' sold in Britain in 2003, contained large amounts of cheap low-grade rice, you can see how slippery the food industry has become.[9]

Now how about you? You are the head of the family. You are sensible about food. You have an allotment. You are healthy, apart from the fact that you suffer from asthma. You eat regular meals at regular times, a good thing from your body's point of view[10.] The processes of digestion and excretion are facilitated if you eat your meals at regular times, and this is partly a matter of conditioning. Just like you dog and cat, you feel hungry at the time that your normal meal is due.

One day you are held up at work, and you realize that your evening meal is going to be delayed. You have a plan for the family meal, but you have not done anything about desert. So on the way home, you pop into a supermarket and buy some grapes. Normally you would buy your fruit at a farmer's market, but these grapes will have to do. You rush home and start work in the kitchen. Soon all the family has assembled. Dinner it late, never mind, we can manage. You all sit down and enjoy your meal as usual, including the grapes produced with an apology. Don't worry - everything is fine. Then you have an asthma attack. The worst you have ever had, so bad that you do not survive. This is our seventh *death by eating*. It happened as a result of eating grapes that had been treated with sulphites to preserve their color. Sulphite sensitivity[11] is particularly common in people who suffer from asthma. You should have known, but you had been brought up on a farm, and you had not been exposed to sulphites before.

Chapter 8   *Food and status*

Anthropological studies tell us that some recent, and presumably ancient, hunter-gatherers lived in a kind of material plenty, and the accumulation of objects, or territory, had not become associated with status[1]. If anything, status was confined to kinship. The young had status as children, the men had status as hunters, and the women as gatherers. Food was distributed by means of reciprocal exchange (see Chapter 5), and was not connected with status, apart from that given to children. Once food is redistributed by someone who is not themselves a food producer, then food and status start to become connected.

Social status can be described as the honor or prestige attached to a person's position in society. In modern societies, status may be determined in a variety of ways; by occupation, ethnic group, gender, religion, and status may or may not be associated with the merits or achievements of the particular person. Historically, however, status has generally been a matter of social stratification. Once the supply of food comes under political control, accounts are kept, and productivity encouraged, then the situation moves towards **stratified redistribution**, with the controllers at the top, the producers in the middle, and the consumers at the bottom (see Chapter 5). In such societies, status refers to the relative rank that an individual holds. This may come about when a person earns their social status by their own achievements (known as achieved status), or when they are placed in the stratification system by their inherited position (called ascribed status).

In pre-modern societies, status differentiation varies widely, ranging from a class or caste systems, to systems based upon merit. Some contemporary empirical sociologists have attempted to produce a single scale of status, usually conceived as a simple index of income, education and occupational prestige (called the socio-economic scale). However, we are not really interested in the mechanics of status, but rather how a person's perception of their own status influences their food preferences.

Income and status

Money can be an indication of status in a contradictory or even inconsistent manner. For example, a drug dealer may acquire a lot of money, but have low social status (i.e. status in other people's eyes), whereas a teacher may earn little money, but acquire high social status. However, drug dealers may have high status within their own (criminal) community, and they may be indifferent to their low status within the larger society. Moreover, there is evidence that people who conceive of themselves as having high status live longer that other people[2], and this suggests that what really has an effect upon a person's food preferences and health is their beliefs about their own status.

Money implies status even if only in the person's own eyes. The advantage of money is that it can be a kind of substitute for land, or for objects such as pictures, or for whatever the holder imagines to be associated with status. So if you are worried about what your friends think of you, you can buy pictures that visitors to your home can admire. If you are worried about what people at large think of you, you can spend the same money on an expensive car.

If your status is ascribed in the sense that you belong to a particular family, or ethnic group, or religion, then your food preferences are likely to be influenced by what you think is expected of you. When a monarch or chieftain, eats in public, then the food will be determined by the (perceived) public expectations. Similarly, if you belong to a particular ethnic group, or religion, then what you eat will be constrained by what is expected of you.

So much depends upon how you see yourself socially. If you identify with a particular social group or class, then you are likely to have ties to that group, and to some extent, to see yourself judged by them. Thus you may have ties to people of the same ethnic group as yourself; ties to others of the same religion, or the same political views, or the same sex or sexuality, or the same income group. In other words, if you have social ties to a particular group, or identify with a particular social group, then you will see other members of the group as your peers.

How does all this affect what you eat? What you eat depends partly upon your upbringing – your status as a child. As we saw in Chapter 7, your genes, and what you learn, affect your food preferences in a direct way. But your upbringing also affects your (self-perceived) status, because you may be born into a family of a particular religion, particular political views, and particular income group. Your upbringing determines your peers, throughout your family life and your schooling. Your peers influence your behavior, and particularly you're eating behavior.

Peer pressure

If you were an important person in ancient Rome, in ancient China, in Medieval Europe, in Viking Scandinavia, or in pre-British India, you would host banquets from time to time. The food provided would be exotic and expensive, way beyond the means of an ordinary person. The food provided is a symbol of your status. These days, when the Queen of England hosts a state banquet, the food provided is probably not beyond the means of many of the guests, but the setting and the scale probably would be. How you set about impressing your guests at a banquet is determined largely by your peers. In any particular period of history, you are influenced by what the other Barons, Moguls or Monarchs are doing. You are influenced by peer pressure.

Even at the present-day humble dinner party there is a certain element of rivalry. You may be returning hospitality, but hospitality is usually returned like-with-like. In inviting people to dinner, there is often a kind of balanced reciprocity. Food offered as part of hospitality promotes good social relationships, including a certain pressure to reciprocate. Thus if someone invites you to dinner, you are under a certain social pressure to invite them back, at some time in the future. And when you do so there is a certain pressure to make an effort to match or better the dinner they gave you. The pressures to conform in situations that involve balanced reciprocity go back a long way – all the way back to our hunter-gatherer ancestors.[1]

Peer pressure involves not only returning hospitality, but also to conforming to the type of food eaten. If you belong to a community, whether it be your extended family, your school, or your religious group, there is often a pressure to conform to a standard notion of acceptability – what is OK to eat, and what is not OK. Sometimes these standards are formalized, as in many religions. Amongst school children in particular, such pressures can be intense. What you eat can become a matter of status amongst your peers.

Let us look at school children who bring their lunch to school, rather than eating the lunch provided by the school. Such children often trade items at lunchtime. This is entirely reasonable – if you feel like a carrot and your friend has one and you do not, then you might offer something in exchange. The problems start when your parents do not allow you to take – say - sweets, but you like sweets and your friend has some

119

sweets. Obviously you will try to trade. But your parents probably have a reason for not allowing sweets. They might be concerned about your teeth, or your bodyweight, or the fact that your grandfather had diabetes. You know about your parents concerns, but you also believe (on the basis of the evidence) that your friend's parents do not have those concerns. You become aware that your friend has a status that you do not have. He has this status because his parents are like most other parents – they allow sweets. You like sweets and you have a tendency to conform to the norm, so you offer to trade. If you were in a school where the (parental) norm was no sweets, then you probably would not want to trade. So there can be a conflict between what your body is telling you (get sugar) and what your brain is telling you (sugar can be bad for you). It is peer pressure that tips the balance in this kind of conflict.

Now when you get home you watch television, including the ads. Because most children go home from school at an earlier time than working adults, the advertising companies (naturally) target their advertisements at children. Various food items are

paraded as being delicious, easily available and cool. What does cool mean? It means not only socially acceptable amongst your peers, but having a certain prestige or cachet. So there you are – not only is it OK amongst your school friends, but it is de rigueur amongst all people of your age. You would not want to miss out, and you particularly would not want to be seen to miss out. Moreover, the child-friendly advertisers are out to help you. They have your interests at heart. They are sugaring the pill – a pill that will affect you for many years.

Now you have grown up somewhat. You are a college student (well done). You have developed certain eating habits and food preferences, and you are concerned about your body.

Body Image

How do you feel about your body? If you feel comfortable and confident in your body, you have a positive body image. If you feel ashamed, self-conscious, and anxious about your body you have a negative body image. Body image is the way you see yourself when you look in the mirror or when you picture yourself in your mind, and what you believe about your own appearance. According to the National Eating Disorders Association, people with negative body image have a greater likelihood of developing an eating disorder and are more likely to suffer from feelings of depression, isolation, low self-esteem, and obsessions with weight loss

Some researchers claim that college females deliberately alter their self-reported body image according to characteristics of their prospective audience. In particular "fat talk" is a way of impressing other women by degrading one's own body image. For example, 'I am too fat and I am determined to lose weight'. If, in fact the speaker is perfectly normal, then the listener is likely to think that she too is too fat. In other words "fat talk" is a form of peer pressure. The researchers claim that this is a social norm amongst female college students in the USA[3].

Other research indicates many men wish to become more muscular than they currently perceive themselves to be[3]. Now it is quite possible that our hunter-gatherer male ancestors were more muscular, or at fitter than we are, simply because they had more exercise. So it does not seem all that surprising that some men are dissatisfied with their bodies. Moreover, the remedy is in their own hands – take more exercise. A perfectly normal woman who claims she is too fat is a different matter. For one thing, there have been times in the past when it was fashionable to be plump, and there have been times when fat women were preferred, because obesity was considered a symbol of wealth and social status, especially in cultures prone to food shortages. So why would a modern women claim to be fatter than she really is? Firstly, the claim may be disingenuous– a social ploy. Secondly, the claim may be genuine in the sense that she would like to be a bit thinner, but is not actually going to do anything about it. Thirdly, she may really believe that she is too fat, and is going to do something about it. In this third case there is a conflict between what her body is telling her (everything is OK) and what her mind is telling her (I am too fat).

There are some men and women who are too fat, and know it. What they know is that they are unhealthily fat and are likely to suffer for it at some time in the future. Whether they want to do anything about it, and whether they are capable of doing anything about it, are questions that we will address later. Here, we are interested in those situations where there is apparently a conflict between mind and body. So leaving aside (for now) the situation where someone believes that they are too fat, or too, thin and are aware of the health hazards, but have not (yet) made an effort to do something about it; what about people who believe (wrongly) that they are too fat (or thin)?

Some people make misjudgments about their body shape or size while looking at themselves naked in a mirror. They are misjudgments, because they do not make the same judgments when they look at other people who are the same size and shape as themselves. If such self-evaluation leads to considerable weight loss due to reduced food consumption, and denial of the seriousness of the current low body weight, then the person is probably suffering from some kind of mental disturbance. Some such people have an intense fear of gaining weight or becoming obese, even though in reality they are underweight. When such people refuse to maintain body weight at or above a minimally normal weight for their age and height, they have a life-threatening condition.

Extreme weight loss usually results in endocrine disorders, resulting in lack of menstruation in women and impotence in men; low blood pressure and heart-beat disorders; immune-system dysfunction; reduced bone-mineral density, and other features of starvation (see Chapters 1 and 10). When these symptoms result from a refusal to eat (enough), then the person concerned may be suffering from *anorexia nervosa,*[4] but note that similar symptoms occur in people on hunger strike (see Chapter 10). Some of these symptoms can also occur in people who put their bodies under considerable physical stress and do not compensate for this with a suitable diet. Some ballet dancers and gymnasts come into this category.

Many anorexics and professional performers may endanger their health (by not having a suitable diet) as a result of social pressure. In the former case the pressure can be said to be imagined, whereas in the latter case it emanates from the competitive environment that dancers and professional gymnasts inhabit. In the one case there is a mind-body conflict, while in the other case there is a conflict of interests (i.e. health vs. professional advancement). Although it is true that there is a possible genetic component in anorexia[4], and that some (e.g. ballet dancers) become anorexic as a result of professional pressures[4], the majority of anorexics are influenced by social aspirations. In 2003 an epidemiological study of 989,871 Swedish residents indicated that the chances of developing anorexia were connected with gender (more common in females), socio-economic status (more common in the wealthy) and ethnicity (people with non-European parents were the least likely to be diagnosed with the condition). These and other similar findings suggest that the body image is perceived, not in isolation, but in relation to a social context[4]

The status of food itself.

In the early 19th century oysters were very cheap and were mainly eaten by the relatively poor. As a food they had a low position, or low status. By the mid 20th century oysters were expensive, eaten only by the rich. They had high status. By the end of the 20th century oysters were widely affordable. The changing status of the oyster is largely due to its availability. When wild oysters were commonplace, they were inexpensive, and associated with ordinary people. As human populations grew, demand for oysters grew, and the oyster beds were depopulated. As the availability of oysters declined, the price rose, and they became associated with the rich and fashionable. Then oysters became more readily available as a result of artificial oyster culture. They started to appear in supermarkets, and in the UK their price has dropped (in relative terms) over the past thirty years. A similar story could be told about salmon. Food has high status when it is scarce and therefore expensive and low status when it is commonplace.

Commonplace food is largely a matter of fashion. For example, Florence White's 'Good Things in England', "A practical cookery book for everyday use, containing traditional and regional recipes suited to modern tastes", was published in 1968. In the 42 pages devoted to breakfasts there is no mention of breakfast cereals or muesli. Breakfast cereal was invented by John Harvey Kellog in 1894, but it was promoted by his brother Will Kellogg. He advertised it as no-one had ever advertised before. He gave away free mini-packets, he realized the importance of the box as a purveyor of messages, he promoted his advertising messages with distinct recognizability and continuous repetition, and he targeted children as well as adults. From the advertising point of view Kellogg did everything right. When the depression came, he did not cut back, but redoubled his production. The result was that many

Americans gave up hot breakfasts, and opted for the fast, easy, cold breakfast[5].

Muesli was developed around 1900 by a Swiss physician for his patients in his hospital. It became popular as a breakfast food in western countries in the 1960s. These two types of cold breakfast foods acquired their status in different ways. Manufactured cereals acquired their position in the market through advertising, while muesli acquired its status largely through grass-roots influence. The former are complex products of the food industry, while the latter can be made at home from readily available ingredients.

If you were a big time food manufacturer, what would you want? You would want your food to be eaten - as much of it as possible. How to achieve this? Firstly, you would make your product as cheap as you can whilst still making a profit. You could do this by using cheap ingredients, and by increasing the scale of the manufacturing process. Secondly, you could advertise you food as widely, effectively, and inexpensively as possible. These two ways and means apply not just to food, but too many aspects of manufacturing. Thirdly you could make your food addictive in the sense that buying it, and eating it, becomes a habit. This approach is widely used in marketing legal products, such as tobacco and alcohol, and also for some illegal products, so why not for food?

The question is – how to do it? How do you go about making your food product habitual? The answer is - make it attractive, make it convenient, and make it habit forming. Making food attractive is much the same as making it high status. Food can be attractive because it looks attractive, because it is fashionable, or because it tastes good. Food can be made convenient by making it fast (meaning no effort on the part of the consumer), by making it easy (as in ready-meals), and by making it handy (i.e. you don't have to put yourself out to get it). This approach trades on people's inherent laziness.

So how do you make your product habit forming? As we saw in Chapter7, animals (including) humans can be subject to *conditioning*. For example, pigeons are easily conditioned to peck at an object to obtain a food reward, but they are not so easily conditioned to tread on an object to obtain the same reward. Pecking to obtain food is natural for a pigeon, but treading is part of courtship, and it is difficult for a pigeon to associate treading with food. So what to humans readily associate with food? They associate certain internal consequences, such as a feeling of fullness (messages from the stomach), and increase in blood glucose (measured in the brain), a change in salt intake (measured by taste, and indirectly by internal changes), and an increase in absorbed fatty acids (indicating energy storage). These signals, outlined in Chapter 1, are basically the reward signals of eating. Thus food can be habit forming when it contains certain ingredients, those ingredients that provide rapid information about the good consequences of food intake, as distinct from the slow and indirect information that comes from the intake of vitamins, minerals, etc (see Chapter 1).

What the food manufacturer should do, to induce eating habits, is to give the food a distinctive visual appearance and flavor, and to load it with sugar, salt and fats. This is exactly what many of them do, especially those making fast food, and ready meals. In this way their customers become conditioned to their product. Buying and eating their product becomes a habit. Thus, many people develop the habit of buying in pizza (you can have it delivered to your door), burgers (there is usually a sales outlet just around the corner), and Chinese and Indian takeaways. They are everywhere. I once lived in a street where there was a takeaway that seemed to be open all night and all day. It was called East Wind Take Away Chinese Restaurant. (I eagerly awaited the day when this would happen, but I was disappointed.)

Fast food has been available for centuries. In medieval Europe there were pie shops and cookhouses in the larger towns. Cornish pasties go back to the 13th century. Such food was available in taverns, and in street markets, while in coastal areas seafood was sold at the quayside, especially eels and oysters. The development of trawler fishing led to the British favorite fish and chips, by the mid 19th century. The main differences between then and now lie in the ingredients and in pattern of consumption. The modern history of fast-food restaurants in the USA can be traced to the 'automat', a shop or restaurant featuring prepared foods behind small glass windows and coin-operated slots. These first appeared in Philadelphia in 1902. They remained popular up to the 1940s. The take-out food was sometimes promoted by the slogan "Less work for Mother". This indicates a shift in the perception of the burden and onus of providing food within the family – a shift from eat-at-home to get-it-yourself.

The contents and consequences of eating, modern fast food have been well documented[6]. It is all relatively simple. If you habitually have too much fat, too much, salt and too much sugar in your diet, you are going to put on weight, and your are likely to have high blood pressure. You may become obese. Not to worry, lots of other people are obese. In fact in the USA reliance on high energy fast-food meals tripled between 1977 and 1995, and calorie intake quadrupled over the same period. Currently, about 64.5%, of US adults are either overweight or obese.[6] However, not all obesity is due solely to overeating. The calorific imbalance that results in obesity is probably the result of a combination of genetic and environmental factors, and various genetic conditions that feature obesity have been identified[6]. Certain ethnic groups may be more prone to obesity than others, because the ability to take advantage of rare periods of abundance by storing energy efficiently may have been an evolutionary advantage in times when food was scarce. Moreover, heavier people tend to form relationships with each

127

other[6], so whole families may carry a gene that promotes fat storage. Nevertheless, there remains a strong correlation between the consumption of fast food and obesity in certain countries, and there is a strong correlation between obesity and certain medical conditions, such as osteoarthritis (your skeleton has to carry more weight), sleep apnea (waking in the night due to lack of oxygen in the blood supply to the brain), and cancer, cardiovascular disease, diabetes, non-alcoholic fatty liver disease (due to the increased number of fat cells). All-in-all, if you are obese you are likely to have a shortened life.

So here you are, a modern city dweller, faced with all sorts of temptations. But you are concerned for your health, and you do something about it. Nothing in this chapter (so far) is news to you. Your status at work is important to you, and you don't want to look fat or lazy. On the other hand, you like your food. You find it hard to resist your favorite foods, but when you have eaten a large meal you start to worry about the consequences. You don't need to worry, there is a way out of this dilemma. You have heard of people who self-induce vomiting after meals, but that does not appeal to you at all. That is bulimia – horrid. What you can do, and what you do, is take laxatives. This means that much of you food goes down the toilet and is not absorbed into the body. It is painless and it works. You can eat and yet look slim. Your status is intact. What you don't realize is that prolonged use of laxatives disrupts your electrolyte balance, because the intestinal fluids are expelled, instead of being reabsorbed in your lower intestine (see Chapter 1). The electrolyte balance is important for proper functioning of the heart, and some people who have your form of bulimia (yes you do have bulimia[7]) are satisfactorily treated in hospital. But you die, because you kept your secret to yourself. You witheld information from your doctor, and you were mistakenly diagnosed with pancreatitis[7]. This is our eighth case of *death by eating*, a death largely due to (perceived) social pressures leading to secret self-treatment.

Chapter 9   *Food panic*

Some people are faddy about their food – they don't like this, they don't like that – or, they insist on this and they insist on that. What is all this about? We have seen that some people have genetic predispositions against certain foods, and some people have had bad experiences in the past that they associate with certain foods. But what about people who insist on eating certain things, and avoiding other things, as a matter of routine? What are these food fads about?

Food Fads

Food fads are based upon beliefs. Before the person puts food into their mouth, they have a belief about the food. They may believe that it will taste unpleasant, or pleasant. They may believe that it is good or bad for their health. They may believe that they ought, or ought not, to eat the food for moral or religious reasons. Moral and religious beliefs have diverse origins. Some people believe that it is wrong to kill animals, so they do not eat animals. Some believe that it is wrong to eat certain animals, so they do not eat those animals. The origins of these beliefs may lie in religious doctrine, or in self-conviction about some aspect of the world we live in today. What we have here are cases of mind over matter, the subject of the next chapter.

So leaving aside beliefs of a moral nature, and beliefs about what the food will taste like, which is basically an aspect of neophobia (see Chapters 1 and 3), we are left with beliefs about health. Fad diets that are believed by their practitioners to improve health are often irrational. In other words, the person concerned does not have good reason, or good evidence, that the diet will improve their health. Why not? – Because such diets are often the result of promotion and advertising that is poorly, or misleadingly, based upon science. When I was an undergraduate, reading biochemistry in the 1950's, it was discovered that some people who died in car accidents, and were then subject to autopsy, had clogged-up arteries. The unfortunate people who died in this way were of all ages, and this enabled researchers to correlate the condition of their blood vessels with their bodyweight, fat deposits, and dietary lifestyle. They found that much of the material impeding the flow of blood in the arteries was made up of cholesterol, and we were taught what cholesterol was and how it was synthesized in the body. A decade later, advertisements started appearing, claiming that the food on sale was low in cholesterol, and therefore healthier than food that was high in cholesterol. I did not believe a word of it, because I had been taught that the cholesterol that one eats is broken down into its basic building blocks during the process of digestion. If a person had cholesterol deposits in their arteries, the cholesterol must have been synthesized within the body. As it turns out, 50 years later, it is now known that a small amount of cholesterol comes directly from your diet, and the majority is produced by your liver. However, if your diet is high in saturated fats and cholesterol this can cause your liver to produce more of a certain type of (harmful) cholesterol[1]. However, we are not so interested in the scientific details (here) as we are in the social aspects.

Scientific 'news' about cholesterol, vitamins, etc. has proved to be a godsend for the food industry. It has given them extra weapons in the fight against rivals. For example, in the war between margarine and butter, we see margarine advertised as 'low cholesterol' disregarding its other strange ingredients[2].

Breakfast cereals, and milk, are advertised as being 'fortified' with various vitamins. The result of all this sudden concern for our health has been confusion and panic amongst ordinary people. Such advertising is fuel for food faddism.

Part of the problem is that many people are ignorant about science. They don't understand how science works, and they can't tell the difference between good science and bad science. In biological terms, they are similar to those birds that can't tell the difference between their own eggs and a cuckoo's egg. The birds are perfectly capable of learning to make the discrimination, but they don't have the time and they can't be bothered. Consequently, advertisers can get away with statements like 'our additives are scientifically proven to have this beneficial effect'. The general public doesn't understand that nothing is scientifically proven, that scientists hold to theories only so long as they fail to be discredited or contradicted by new evidence. In some cases this takes a long time (even centuries), and in other cases it takes a short time (even a few days). So there develops a mythology about certain types of food and certain types of diet. This is fertile territory for those who see the opportunity to make money. In evolutionary terms, it is fertile territory for predators and for parasites.

If you regularly include a particular type of food in your diet, because you think it will help to cure a specific disease, or if you eliminate a normal food from your diet because you think it is harmful, or if you eat special food to express a particular lifestyle, then you are probably a food faddist. Belief in fad diets is often irrational and idiosyncratic, and many individuals who adhere to fad diets will not consider recommendations made by dieticians and nutritionists[3] Extreme diets often lack sufficient calories, essential vitamins, suitable protein, and some minerals that are essential for growing children. Parents forcing children to adhere to fad diets to the point of severe nutritional disorders is considered, by experts, to be a form of child abuse[3] .Nevertheless, there are people who have an unshakable belief in the efficacy of their chosen diet. How can this be?

Firstly, many forms of food fetish are supported by pseudo-scientific claims. Such diets may claim to be scientific, but if they do not follow the scientific method in establishing their validity. For example, I drink a lot of green tea because I think it is good for my health. I think this because I have read some of the scientific literature. Now if I discovered a scientific study that claimed that green tea was bad for me in some way, then I would investigate the matter. If I became convinced that the new study carried scientific weight, in the sense that the evidence and the conclusions form the evidence, added up, then I would stop drinking green tea. As mentioned above, real science is always open to revision. There may be observations that prompt particular explanations, but taken alone, these should not be used as evidence of the validity of the explanation.[3] Moreover, fad diet promoters often ignore or refute what is known about fundamental associations between food intake and human health[3]. For example, there is no good evidence that weight loss can be enhanced by any diet, apart from one involving caloric restriction[3]. Many fad promoters claim that their diet produces sustainable weight loss, but the evidence is lacking.

Secondly, fad promoters may suggest spiritual, moral, or religious reasons in support of a particular diet. Some of these promotions may be sincere, but it is quite clear (from the advertisements), that some are spurious. Sincere advocates, of what we might call doctrinaire prescriptions, may base their approach on holy writ; on tradition; on the belief that there is a moral obligation to eat, or not eat certain foods; or simply on the belief that the discipline involved in the fad diet is a good thing in itself. Insincere fad promoters are primarily interested in making money, exercising power, or establishing some sort of prestige.

Thirdly, some fads can be justified on scientific grounds. For example, it has long been known that peoples living in the arctic tundra do not have a diet rich in fruit and vegetables, but neither do they seem to suffer from dietary deficiencies. In 1944 scientist Hugh Sinclair noticed that the indigenous people of northern Canada, the Inuit, did not appear to suffer from lack of vitamin A, or vitamin D, nor did they seem to suffer from heart attacks, despite a diet high in saturated fat. In 1976, together with two Danish scientists, he studied Eskimos in Greenland, and found that their blood cholesterol levels were no different from our own. Maybe they had evolved some kind of resistance to the bad effects of a high fat diet. To test this possibility, three years later, he put himself on an Eskimo diet. He imported a whole seal and lived of this, plus fish and shell fish – a very restricted diet. Within three months he had lost 28 pounds, and his blood fat composition was the same as that of the Eskimos. Also, the clotting ability of his blood was greatly impaired, reducing the risk of internal blood clots, a major cause of strokes and heart attacks. Clearly, there was something in the diet that protecting the indigenous people from these health hazards.

It was subsequently found that the same results can be obtained from cod liver oil. In fact cod liver oil had been known for a long time as a source of vitamin D, but it also contains active vitamin A which can be dangerous in large doses. Fish oils extracted without the liver component are free of this hazard. They are taken from fish such as mackerel, tuna and salmon.

These fish prey upon sardines and herring, and these in turn feed on micro algae that produce omega-3 fatty acids, the protective substance that Sinclair was looking for. Unfortunately, these days, the fish at the top of the food chain

133

tend to accumulate toxic contaminants such as mercury, dioxin, chlordane, etc., all by-products of human industry.[4]. Fish oil, derived from the tissues of these fish, namely mackerel and tuna, has a number of beneficial health effects, provided that it is purified. The beneficial health effects of omega-3 fatty acids have been the subject of much scientific research[4]. Taking food supplements containing omega-3 is a food fad based upon a belief that it is good for your health. In this case the belief is justified by evidence, unlike some other food fads.

Dieting

Going on a diet? This expression amuses me, because the person concerned is on a diet already. Your diet is your habitual food. What they mean is going on a different diet. Why would anyone want to change their diet? Lots of reasons – to lose weight, to gain weight, to find which component of the diet they are allergic to, and many more. A person deliberately changes their diet because they believe that the new diet will benefit them in some way. They may hold such beliefs as a result of receiving medical advice, as a result of advertisements, or as a result of information gained on the grape vine.

Some weight-loss diets restrict the intake of food in general, or of specific foods, with the aim of reducing body weight. People differ in their lifestyle and metabolism, so what works for one person may not work for another. Moreover, many dieters fail to maintain significant weight loss over time. Amongst those that have lost more than 10% of their bodyweight, only 20% are able to maintain the new weight for a full year.[5] Weight loss typically involves the loss of fat, muscle and water, and different diets will have different effects of these components of the body.

The most effective way to lose weight is to lose fat, rather than muscle, because muscles do work and fat does not. So the more muscle mass one has, the higher one's metabolic rate is[6], and the more calories are expended. Moreover muscles are denser than fat, so fat loss (per unit weight loss) results in a greater reduction in physical bulk. Muscle loss can be restricted by maintaining sufficient protein intake, and by muscular exercise, during a period of weight loss. A person on a low carbohydrate diet, while doing strenuous exercise, should increase their protein intake, but not too much, because excessive protein intake can exacerbate liver and kidney problems especially in individuals who already have such problems.[5].

All bodily processes require energy. During exercise, when the body is expending more energy than it is taking in, the body relies on internally stored energy sources (see Chapter 1). The first source the body turns to is glycogen stored in skeletal muscles and the liver. When the glycogen stores are nearly depleted, the body begins lipolysis, the mobilization of fat stores. When both carbohydrate and fat stores are depleted, the body will obtain energy from muscle, which is harmful for the body. When losing weight you must be careful not to lose muscle.

The best way to lose weight is to reduce the amount of carbohydrate and fat in the diet, and to increase the amount of exercise. However, to be effective, a considerable amount of exercise is necessary. Exercising three times a week and raising your heart rate is good for the heart, but the effect on weight loss will be negligible. One problem is that unfit people become tired after a small amount of exercise. If you do an hour's good exercise per day for a month you will begin to lose weight, provided you do not increase the amount that you eat. But the increased exercise will probably make you hungry, and if you respond to this by eating more, then you will be back to square one. The thing to do is to gradually increase your exercise and restrict your diet, so giving your body time to adjust. But no doubt you are impatient.

You can see that someone in this kind of situation is an easy target for those promoting quick solutions. Some advocate the use of drugs to control appetite, but this can be dangerous. For example, diuretics (in either herbal or in pharmaceutical form) effectively induce weight loss via the excretion of water (i.e. water is taken from the blood by the kidneys and excreted as urine), but they have no effect on body fat. Diuretics can cause thickening of the blood (so that the heart has to work harder) and can damage the kidneys and liver. Other drugs, such as amphetamines and ephedrine, have been banned in many countries. On the whole, interfering with the regulatory systems of the body is not a good idea (except in situations of medical emergency[7]). This is especially true for children. A well-balanced diet is critical during childhood and adolescence. Many specialty diets are unhealthy because they deprive the body of essential nutritive building blocks. Some studies have found that dieting is counterproductive, because children who diet are prone to binge eating, and so gain more, not less, weight than non-dieters. Their bodies are telling them something, and they respond to those messages.

One message that your body sends your brain comes from your stomach. As your stomach fills up, *satiety* signals are sent to the brain. These relate to the volume of food in the stomach. Of course, these are not the only messages about the consequences of eating. First of all you see the food, then you taste it, then you swallow it, then you get messages from the stomach, then you get more internal messages as the food is processed. Once the products of digestion start to enter the bloodstream, messages about changes in the composition of the blood are sent to the brain (see Chapter 1). Usually, you will have stopped eating before these messages arrive. You will have stopped eating partly as a result of messages from the

136

stomach, and partly as a result of seeing how much food you have eaten. In fact, you can often predict how much food you will eat, before you even start.

Consider the following situation. You are in a cafeteria, where you can see and pick up the food that you want to eat, as you move along the queue towards the till. You cannot taste the food before you have paid for it, but you are quite capable of judging how much you are going to pay for and how much you are going to eat. In other words, you can predict how much you will eat on the basis of your hunger and the sight of the food. When you think about it, this is quite a feat, because the food comes in many guises. Sausage rolls contain carbohydrate, protein and fat, bread rolls contain carbohydrate, and chocolate contains mainly fat and sugar, and so on. On the basis of your past experience, you can make the complicated judgment, as to what and how much food you should buy. If you buy too little you will remain hungry; if you buy too much you will waste money, but actually you get it right.

Physiologist Michel Cabanac has carried out experiments on this type of situation. In one set of experiments, ten young males were invited "to have lunch in the laboratory at their own usual time for lunch, and receive $12." Each subject came on four different days, one preliminary session and three experimental sessions. In the first session they were asked to eat one small (one mouthful) sandwich from each of ten plates, and to give a magnitude-estimate verbal rating of the pleasantness (positive rating) or unpleasantness (negative rating) of each item. The rating was to be a number of their own choice. After first eating ten different sandwiches, the subject ate another one from each plate to modify, if necessary, the rating they had given to each type of sandwich. The subject was then told that he could eat whatever sandwiches were left, because he had been promised a meal. The total number of sandwiches eaten in this preliminary session by each subject was recorded.

In the following experimental sessions, the price of each type of sandwich was set according to each subjects rating in the preliminary session. The subjects had to pay for sandwiches that they had rated positive, and were paid for eating sandwiches that they had rated negative. The rates of pay varied from one session to another, there being little difference between sandwich types in the first experimental session, medium differences in the second session, and large differences in the third session. Subjects were told to eat the same number of sandwiches that they had eaten in the preliminary session. The subjects paid for their food out of the $12 that they were given for attending each session.

The results showed that not all subjects preferred the same sandwiches. However the price rank-order of the ten types of sandwich presented to each subject was based upon their preferences as revealed in the preliminary session. The results of the first experimental session showed that the subjects ate mostly medium and high-palatability sandwiches, but no low-palatability sandwiches, although they could have received a small amount of money for eating the latter. In the second experimental session the subjects ate sandwiches from all the plates. In the third session, they ate sandwiches from all the plates, but the majority was taken from the low-palatability plates.

These results were entirely consistent with the mathematical predictions made on the basis of the pricing of the sandwiches in the three experimental sessions. The hypothesis underlying these predictions was that the subjects should eat more sandwiches whose palatability outweighed the price than sandwiches whose palatability was not worth the price.[8] In other words the subjects could trade-off money expenditure vs palatability in the same way that animals trade-off energy expenditure vs returns on foraging (see Chapter 3). Moreover, the subjects were consistent and accurate, just like you would be in the cafeteria situation.

So knowing when to stop eating is a complex matter, involving the sight of food (and what you have learned about that food), the price of food (and what you have learned about value-for-money), the taste of food (and what you have learned about that), the short-term satiety signals (from the stomach), and the post-digestive consequences of eating (see Chapter 1). All these factors contribute to your eating habits. So what about the person who wishes to change their diet to lose weight? They should eat less and exercise more – but how? It is not that easy to change your eating habits. As we shall see, there is no-end of people willing to advise you, at a price. You do not need to pay this price if you understand the science. So what would a scientist say?

Nutrition Professor Barbara Rolls is a scientist who has spent her working life studying eating and drinking. Her advice is simple – people who take in more calories than they expend put on weight. So what you should do is think about the energy density of your food. The energy density of a food item is the amount of energy stored in the food per unit volume, or per unit weight. This is the energy released when the food is metabolized by a healthy organism when it ingests the food. Thus typical values of food energy value of a typical hamburger might be 2.5 kilocalories per gram (i.e. 2500 calories/g). Barbara Rolls' approach is to help people find foods that they can eat, and feel full, while still losing weight. People feel full (in the short run) because of the amount of food they eat - not because of the number of calories. Therefore the thing to do is to eat low energy-density food. Low density food includes non-starchy vegetables, low-fat milk, soups and broths. These are low density, because they contain a high proportion of water (the energy density of water is zero). Foods with high fiber content are also low density, because fiber (cellulose) is not digested by humans.[9]

Barbara Rolls' approach, called volumetrics, does not ban food types, or divide foods into the good and the bad. It advises people to choose foods on the basis of their energy density, whilst maintaining a balanced diet. So if you choose to eat grapes rather than raisins, you are taking the low energy density option, and this gives you a much bigger portion of foods for the same number of calories, and can help you to feel full. Her approach essentially boils down to the sensible diet that any nutritionist would recommend - lower-calories, lower-fat, with lots of vegetables and fruits – not unlike the Paleolithic diet.

Panic, parasites and predators

In biological terms, parasitism is an interaction between members of different species, in which one (the parasite) exploits the other (the host), but generally does not kill it. Many parasites are fully dependent upon their host at some stage of their life cycle, and natural selection therefore exerts a strong pressure for them to become highly adapted to their niche, and not destroy it. At the same time, many host species have evolved defensive measures, so there is often an evolutionary 'arms race' between parasite and host.

In terms of the evolution of human food, parasitism is an interaction between members of different economic players (e.g. the producer and the consumer), in which one (the parasite) exploits the other (the host), but generally does not kill it. . Many parasites (e.g. food corporations) are fully dependent upon their host (e.g. the consumer), and market forces therefore exerts a strong pressure for them to become highly adapted to their niche, and not to destroy it. At the same time, many hosts (i.e. consumers) have developed defensive measures (i.e. they shop elsewhere), so there is often an evolutionary 'arms race' between parasite and host.

The food corporates, and their allies the supermarkets, attempt (with some success) to influence the preferences of consumers (by advertising and spiking their products with must-have ingredients), and the consumers (or some of them) rumble them, and take their business elsewhere. Moreover, the consumers have votes, and their governments (eventually) intervene to restrict the most outrageous deceits on the part of the parasitic corporates. Nevertheless, it is clear from the evidence (the growth of obesity, the increasing prevalence of allergies, and the increasing psychological impact of fast food) that the 'arms race' is currently being won by the parasites.

It is not as though the general public were ignorant. We are bombarded with messages on television, in doctors' surgeries, and from government advertising. Awareness is high but participation is low. Why is this? One reason is that the corporates spend much more money on advertising that their opposition, but there is also another reason – food panic.

In Chapter 6 we saw how, in the commercial world, as in the biological world, advertising can be a form of manipulation. The cuckoo, by clever timing and an attractive product, manipulates a bird of another species to brood the cuckoo's eggs and raise the cuckoo's young. Similarly, food and drink manufacturers, by clever timing of advertisements on TV, and an attractive product, manipulate some consumers (against their best interests) to buy and consume the product. Let us take alcopops as an example.

An alcopop is a type of flavored alcoholic beverage, somewhat like a soft drink with added alcohol. They are targeted at the younger drinker, being brightly colored, sweet, and alcoholic. Quite apart from the dubious artificial coloring and other additives, alcopops often contain two substances likely to encourage addiction, namely sugar and alcohol. Research suggests that alcohol addiction, or dependence, is roughly 50/50 genetic/environmental.[10] Either way, encouraging young people to consume alcohol under a misleading guise is likely to have a beneficial effect upon sales, both in the short term and in the longer term. In Chapter 7 we saw that sugar is highly habit forming. Some scientists claim that, for some people, sugar is truly addictive. Other scientists prefer to refer to sugar dependence. In 2003 the World Health Organization and the Food and Agriculture Organization, issued a report compiled by a panel of 30 international experts. It stated that sugar should not account for more than 10% of a healthy diet.[10] Combining alcohol with sugar in a well-advertised trendy drink is a good way to encourage addiction, and get your customer hooked. We have seen (Chapter 6) how some parasites manipulate their hosts by brainwashing. Are we seeing it again?

Alcopops appeared on the US market in the late 1980, marketed initially by rum or vodka manufacturers. North American and European governments have been alarmed by this development, but most have responded merely by raising the tax on such drinks. Many consumer protection groups have voiced concerns claiming that alcopops are designed to appeal to younger people, and contribute to underage and binge drinking. One survey found that a quarter of 18-19-year-olds thought there was no alcohol in the alcopops they tasted[10]. So has this attempt to manipulate young peoples drinking habits been rumbled? In a way yes, and in a way no.

The fact that consumer protection groups have protested, and that there have been government enquiries in some countries, might suggest that the public's eyes have been opened to an undesirable market ploy. But sales of alcopops have rocketed, and as might have been predicted, another bout of food panic has been triggered. But, what is food panic?

Food panic

There are three types of food panic: (1) the panic buying of food when prices rise, or there is a threat to supply; (2) the panic induced by publicity about particular foods. Well known examples of the latter include the scare about salmonella in chicken (UK cir 1988), and the scare about CJD contaminated beef (UK cir 1996). These huge surges of food panic seriously affected the market, but they were not entirely unjustified. After all, the governments at the time thought that there was cause for concern. Finally, (3) there is panic that results from rumor. You may not think there is such a thing as this type of food panic –

" But ask yourself why millions of Americans continue to consume foods lacking nutrient value and laden with poisons, smoke tobacco products laced with chemicals designed to increase their addictive potential, watch insipid and enervating TV shows that sap their intelligence and transform their children into zombies, and then, finally, demand that their health care practitioners produce a magic pill to make it all better again, or at least, to dull their mind so they may temporarily forget their problems."[11]

There is little doubt that many western urban people are confused about food. They are bombarded by contradictory messages. On the one hand the food-manufacturing giants, with huge advertising budgets, promote junk food by appealing to status (it is trendy, especially amongst the young), its appearance (but you do not always get what you see in the ad), and its convenience (especially to those who think food preparation is a waste of time). On the other hand, many have

143

realized that junk food is not good for their health, and have switched to alternative health foods. Unfortunately this move is often out of the frying pan into the fire.

"misinformation and disinformation in the popular health literature have robbed people of their common sense and substituted it with a jumble of pseudo-scientific slogans and half-truths" [11]

While the food-manufacturing giants caste their nets, and entrap a large proportion of the population, other predators (sometimes side-branches of the food-giants themselves) are waiting for those escaping the net. Food scares spread by rumor on the grape-vine, are often picked up by the press and exaggerated.

For example, the butter-margarine war (see above) was fuelled largely by the cholesterol scare.

"The major lesson that we should learn from the margarine fiasco is that not paying attention to one's senses can lead to trouble. It is not an accident of nature that many poisonous and harmful plants and substances taste repulsive. Unfortunately, scientists in the food processing industry have studied and designed flavor compounds to mask the unhealthy taste of many synthetic foods, so our task is now more difficult."[11]

So far, in this book, we have traced the evolution of human food from a pre-agricultural situation characterized by nomadic peoples, through settled agricultural communities, to large urban populations, now making up more than 50% of mankind.

"Evolution of human populations together with the accumulated experience of our ancestors has resulted in culinary and herbal traditions uniquely adapted to a population's genetic makeup, local climate, and other factors.

To intelligently prescribe changes based on biochemical findings often requires more extensive knowledge than we currently have. Every year it seems that scientific understanding of nutrition and of vitamin and mineral metabolism becomes more complex, and formerly simplistic ideas about recommended nutritional standards have to be discarded."[11]

One problem is that some present-day populations, such as those of North America, have a very mixed genetic makeup, no longer adapted to the local climate. Even those who are in touch with their family traditions may find them inappropriate in an urban, or sub-urban, situation. Nevertheless, it remains true that many are ignorant basic food biology, and rely on experts, and those masquerading as experts. The result has been a progressive rise in the diseases of affluence. There is a section of the public that is alarmed about this situation, so much so that there is even a very successful website called Quackwatch [12]

Another problem is the mythology concerning healthy eating. The food industry and the health industry are at war. The generals of the food industry have huge resources of ammunition in the form of scientific laboratories and advertising machinery, but their foot-soldiers are largely apathetic. The health industry has less ammunition of this type, but its foot-soldiers tend to be messianic. The result is a plethora of mythology about healthy eating. So how do you manage to navigate through this alligator-infested swamp? You are sensible. You are an office worker, and you do not get out much. You are library pale, but you are aware of the importance of vitamin D. Because you do not get much exposure to the sun, you recognize that a vitamin supplement would be a good idea. But you also know that cod liver oil is probably not the best solution (see above), so you opt for milk that is fortified with vitamin D. You drink a lot of milk over a long period, because you know that milk contains a lot of

145

calcium, and is good for your bones. One day, you start to feel unwell while on holiday. You are disinclined to go to a doctor, because you know you are basically healthy, so you wait until you get home. It is too late – you have acute kidney failure due to over-calcification (an effect of large doses of vitamin D[13]). you die from vitamin poisoning. You had a good diet, you did all the right things, but you were over-enthusiastic. What a shame. This is our ninth case of *death by eating*.

Chapter 10   *Mind over matter*

Our bodies have not changed over the past 50 thousand years, but the food that we put into them has changed in the last 10 thousand years and the pace of change has accelerated recently. Our bodies are basically adapted to a hunter-gatherer diet, but 10,000 years ago our behavior started to change, and our behavior changed our food. In other words, over the past 10000 years our food, part of our ***extended phenotype***, has evolved.

Our food has evolved partly as a result of changes in the environment that are the result of our own behavior. We have progressively burnt down forests, to clear land for agriculture. In some places we have caused erosion and desertification of the landscape. In other places we have flooded the landscape, and in some we have covered it with concrete. Today, some fifty percent of humans are urban dwellers, and it is difficult for them to carry out a hunter-gatherer way of life, or even an agricultural way of life.

So far we humans have managed to adapt to these changes. There have been some genetic changes over the past 10000 years, and there have been many social changes. Our social behavior has evolved from an egalitarian situation to a highly stratified situation. To a large extent, barter has replaced hospitality, and money has replaced barter. Money has been used as a tool to establish slavery, and slavery has been used as a tool to build money-making machines. As we shall discover in this chapter, these machines dictate the agenda these days. They are new players in our evolutionary history. How has all this been possible? In early-human times our bodies ruled. We listened to our bodies and they told us what to do. Nowadays we tell our bodies what to do – it is a case of mind over matter.

Our bodies

When we put food into our bodies, our bodies tell us about it. We see it, we taste it, we chew it, we swallow it, and it arrives in our stomach. If our stomach does not like it, we vomit. If our stomach gives the food the nod, it passes into our intestine. It is now too late to eject the food, but if it causes trouble it can be passed on, and passed out, fairly quickly. The body tells us, in no uncertain terms, that a mistake was made further up the line. We learn from these mistakes.

The food in the stomach is digested, and the products of digestion pass into the bloodstream. The blood circulates throughout the body and the brain, and the brain obtains messages from the blood. The messages can be good, bad, or indifferent. On the basis of these messages we get instructions about what to do about the food we have been eating. The body is in control, as in most animals, but in humans the mind sometimes has ideas of its own.

In Chapter 1 we saw how animals how animals can learn to avoid poisons and to change their behavior if they suffer a dietary deficiency. These mechanisms also exist in humans, but they are sometimes overridden. They are overridden by mental processes, based upon beliefs.

We have seen how beliefs related to our self-perceived social status can influence our eating habits (Chapter 8). We have seen (Chapter 9) how beliefs about our own bodies can dictate our diet, beliefs that are not always well-founded. Nevertheless, it is clear that some people are capable of overriding the messages that their body is sending them. An extreme example is hunger strike.

Hunger strike is a form of passive resistance, or protest, in which participants refuse to eat. Most hunger strikers will take water but not solid or liquid food. The aim of hunger strikers is primarily to gain publicity for a cause, whether it be religious, political or moral. Fasting, as a form of protest has a long history. It occurred in ancient India, and in early Christian Ireland. In more recent times it was a form of protest against British rule in India, it was used by the British and American suffragettes, and by Irish Republicans.

During hunger strike the body obtains energy from stored glycogen for the first three days and from fat stores for about three weeks. After this the body obtains energy from the breakdown of muscle, and other body organs. After 50-70 days without food, the person dies, and a number of people have died in this way[1]. There are three things to say, from an evolutionary viewpoint. Firstly, the person on hunger strike must experience acute hunger symptoms, but these messages from the body are ignored, or overridden by the determination to continue fasting. Here we have a situation of mind over matter, a situation that would not occur in other animals. Most starving animals would be highly motivated to find food. Secondly, if some people can find the will to starve themselves to death, how is it that other people cannot find the will to eat less and lose weight? Thirdly, what is it about human nature that enables us to override signals from the body, signals that have a long evolutionary history and function? To come to grips with these issues, we need to pay more attention to biological function.

Biological function

The function of a biological trait is the increase in reproductive success that the trait confers on its possessor's genetic constitution, or *genotype*. For example, the function of nest-building is to provide a nest to keep the occupants warm, and to help protect them from predation. This implies that these consequences of nest-building confer greater future representation for the genotype of the individuals which practice it.

In most cases, traits that increase the reproductive success of the genotype which produce them also increase the success of their individual carriers, but this is not always true. For example, genes which cause their carriers to assist other individuals sharing the same genotype (i.e. relatives) may be favored by natural selection and may spread through the population, even if they lower the reproductive success of their carriers[2].

Biological functions are always complex and usually multiple. For example, the function of incubation in herring gulls is dependent upon a number of consequences of incubation (see Chapter 1). Thus incubation keeps the eggs warm, but it prevents the sitter from foraging for food. It protects the eggs from predation, but it exposes the adult to attack by larger predators. Thus incubation has both costs and benefits. The function of incubation, therefore, has to be seen in terms of all the consequences of incubation that affect reproductive success, in comparison with the consequences of not incubating, or of incubating in a different manner.

The function of foraging behavior is to provide the individual with energy, and nutrients (see Chapter 3). When the body runs low on energy it sends signals to the brain, and the function of these signals is to organize food-finding behavior. Of course, the brain may have other priorities, and foraging behavior may be postponed. In most animals, the hunger signals, and the ensuing motivation to find food have a genetic basis. In other words, the foraging behavior has *survival value*, and natural selection favors those animals that give high priority to staying alive long enough to reproduce.

Natural selection acts upon the *phenotype* (bodily features and behavior) of an individual. Its effectiveness in changing the nature of a population depends upon the degree to which the phenotypic character is controlled genetically. So, if you learn a foreign language, that ability becomes part of your phenotype. But your children cannot inherit the language (they have to learn it for themselves), so this aspect of you phenotype does not exert any genetic influence upon the population as a whole. Moreover, an individual that has no offspring during its lifetime will exert no genetic influence upon the population (apart from help that it may give to its relatives), however great its ability to survive in the natural environment.

So we have two separate concepts to consider – *survival value* and *reproductive success*. These two are combined in the concept of *fitness*, a measure of the ability of genetic material to perpetuate itself during the course of evolution. Fitness thus depends, not only on the ability of the individual to survive, but also upon the animal's rate of reproduction and the viability of its offspring. The concept of fitness can be applied to individual genes by considering the survival of a particular gene in the gene pool from one generation to another. A gene that can enhance the reproductive success of the animal carrying it will thereby increase its representation in the gene pool. It could do this by influencing the animal's morphology, making it more likely to survive climatic and other hazards, or by influencing its behavior and making it more successful in mating or in raising young. A gene that influences parental behavior will probably be represented in some of the offspring, so that by facilitating parental care the gene itself is more likely to appear in other individuals. Indeed a situation could arise in which the gene could have a deleterious effect upon the animal

151

carrying it, but increase its probability of survival in the offspring. An obvious example is a gene that leads the parent to endanger its own life in attempts to preserve the lives of its progeny.

For most animals, there are certain imperatives that have become genetically embedded during the course of their evolution. These include obtaining food, mating, and parental care. Among humans, however, the situation is more complex. For example, people can, and some do, have fewer offspring than is the norm. They do this as a result of holding certain beliefs. For example, they may believe that their offspring will be viable (or have a good life) only if there are few of them. They may believe that the world is becoming overpopulated, and that it is their moral duty to have few offspring, or they may believe that the world is currently too dangerous for children. The same principles apply to food. During the Neolithic revolution many people obviously thought that growing your own food was a good idea. In fact this move had the effect of shortening the life span, but it also led to a population increase, because it enabled more children to be raised within the family (see Chapter 4). So the new foods reduced survival but increased reproductive success. Now, the idea behind the *extended phenotype*[3] is that the material products of animal behavior, such as the caddis fly house, or the bower bird's bower, are products of the genetics of the individual, and if these artifacts increase the reproductive success of the animal, then the genes involved will be passed to the next generation. Early humans had genes that enabled them to share knowledge and so to grasp agricultural opportunities, and the reproductive success of those that did so was enhanced.

The Neolithic revolution also had evolutionary consequences, because it changed the environment. I used to live near an orchard, about one acre in size, and in the orchard lived a pony. Most of the time the pony was alone in the orchard and it walked round and round, always following the same route. In fact, it had worn a path around the orchard. Why did the pony always follow the same path? Initially it walked its preferred route around the orchard, but as it walked around it began to wear a path, and that made it an even better route, because there were no longer stones and brambles along the path. The pony had changed its environment in a way that, in turn, determined its behavior. This type of phenomenon is fairly common in animal behavior. Altering the environment in such a way as to influence future behavior is called stigmergy. It is the basic principle behind termite mounds, wasp's nests and other examples of animal architecture[3]. In some cases the environmental modification is passed from one generation to the next. Some deer trails in Britain and Germany are known to have been used for centuries, the young learning from the old, generation after generation. This is an example of cultural transmission of behaviour.[4]

As human communities became settled, they changed their environment in such a way that it was no longer possible to return to the previous hunter-gatherer existence. The animals and plants that they had previously relied upon for food became scarce. There was no going back. So what about urban people, can they go back? Urban dwellers cannot grow food in sufficient quantities to feed themselves. There was a time when small towns could be supplied directly from the countryside, but where cities have grown, this is no longer possible. Urban people rely on food manufacturers and distributors. Food distribution and storage costs money. When there is a shortage of food, as during wartime, (or simply because not enough food is being produced), prices rise and the urban populace is the firstto suffer. People in the countryside do have the option of growing their own.

The evolution of human food is affected, not only by physical changes in the environment, but also by social factors. Huge social changes have occurred in the past 10000 years that directly affect the food we eat. There has been a gradual sift from self-sufficiency to dependence upon others. This means that there has been a shift in the evolution of our food. For hunter-gatherers, their homes, their weapons, their cooking materials, and their food, are part of their extended phenotype in a fairly direct way, similar to that of other animals.[3]. They pay for their food with their own energy. For the early farmers, the situation was much the same. But, when people come to rely on others for food, they have to pay for their food indirectly, with money, or by becoming a slave or a serf. Money is a tool with which you can obtain food. It has replaced the spear and the bow and arrow.

Who now decides what food you eat, the supplier or you? Obviously you cannot eat food that is not supplied, so your choices are determined by the supplier. If you are lucky, there will be competing suppliers whose produce is available to you, so you may have some choice. History tells us, however, that there have been certain places, and certain times, when the food choice has been extremely limited. If you are in such a situation, then part of your extended phenotype has been hijacked. How has this happened?

The invaders

Your body, like most animal bodies, is repeatedly invaded by other organisms. Mosquitoes want your blood, tapeworms want to inhabit your intestines and steal your food, etc. Your immune system is a collection of mechanisms that protects against invasion by identifying and foreign bodies. It detects a wide variety of invaders, from viruses to parasitic worms, and distinguishes them from your own healthy cells and tissues. Detection is complicated because invaders adapt and evolve new ways of infecting you. Innate immune systems are found in all plants and animals. Vertebrates have a second line of defense called the adaptive immune system, which is activated

154

when the innate response fails to repel an invasion. The vertebrate immune system adapts its response to improve its recognition of the invader. This improved response is then retained, in the form of an immunological memory, after the invader has been eliminated. This allows the adaptive immune system to mount faster and stronger attacks each time this invader is encountered.[5] However, things can go wrong with your immune system for a variety of reasons, and these may result in allergies.[6]

Risk factors for allergies are partly genetic. Allergic diseases run in families, and identical twins have a 70% chance of having the same allergic reactions, while non-identical twins have a 40% chance of the same. Allergic parents are more likely to have allergic children, and their children's' allergies are likely to be more severe than those from non-allergic parents. Proneness to specific types of allergic reaction is also correlated with gender and with race. Environment is also a major risk factor for allergies. Allergic diseases are more common in industrialized countries than in countries that have a more traditional or agricultural way of life. Within a country there is usually a higher rate of allergic disease in urban populations compared with rural populations. These differences are thought to be due partly to pollution and dietary differences between the two types of environment.[7]

Many allergen related diseases have increased in the western world over the past few decades. This is too short a time for a genetic change in the population and environmental or lifestyle changes are suspected as being responsible. These include increased exposure to allergens due to housing changes and increasing time spent indoors. Moreover, it has been suggested that changes in cleanliness and personal hygiene that have resulted in the decreased activation of a common immune control mechanism. This hypothesis maintains that high living standards and hygienic conditions exposes children to fewer infections and that they, therefore, acquire less immunity. Some evidence suggests that children who grow up on a farm, or keep pets, have a reduced risk of allergy, because they are exposed to food and fecal pathogens, and develop immunity to them. Children denied these exposures are more likely to have asthma and other allergic reactions.[7]

So you can see that there has been evolutionary war between vertebrate animals and their invaders. Some modern lifestyles put people at a disadvantage in that war because they do not develop a strong immune system, but this is not the only war that we humans are fighting. There is now a new type of invader, targeted not at the body, but at the mind. Human beings have not yet developed good mental immunity, and so they are easy meat for this new type of invader.

These new invaders are called corporates, or corporations. A corporation, in modern parlance, is a body of persons who are (collectively) a single legal entity. Corporates have a long history. Their precursors existed ancient India and in ancient Rome. The oldest commercial corporation was probably the Stora Kopparberg mining community. It obtained a charter from the King of Sweden, Magnus Eriksson in 1347. The first modern corporation was the Dutch East India Company, which acted under a charter sanctioned by the Dutch monarch in the seventeenth century.

The main features of a modern corporation are that it has a separate legal personality; is owned by investors (e.g. shareholders); has limited liability; and transferable shares and delegated management (e.g. a board of directors).The defining feature of a corporation is its legal independence from the people who created it. However, corporations can be held legally responsible for criminal offences, such as fraud, and for human rights violations. A corporation dies when it becomes insolvent. When this happens, the shareholders lose their money, and the employees their jobs, but they are not liable for any remaining debts. Modern corporations can be very large. The economic size of a country is usually measured in terms of its gross national product. The twenty largest countries in the world are larger than any corporation but, using an equivalent measure, the next twenty countries (added together) are smaller than the twenty largest corporations (added together). In other words, some corporations are bigger, in economic terms, than many countries. Moreover, many corporations have become transnational or multinational, a process called globalization. Consequently some have attained positions of considerable power. These corporations have grown by acquisition, or by merging with each other. Just as the human species spread all over the world, supplanting and displacing other species, so this new kind of organism is repeating the process.

Corporates are a new kind of invader, and they invade in two ways – by altering the environment, and by altering your mind. Leaving aside the environmental damage caused by mining, oil installations, etc, let us concentrate on food. There was a time when Africans (for example) were self-sufficient. Now they are not. Why?

- overpopulation? somewhat
- climate change? somewhat
- political instability? somewhat
- corporate interference? indeed

So what are the consequences?

- movement of people from the countryside (where they can grow food) to the towns (where they cannot grow food)
- in the countryside, cash crops are grown instead of food.

- In the towns, people are encouraged to eat and drink stuff produced by corporates
- In poor countries a gap develops, which aid agencies attempt to fill
- so people come to rely on aid agencies
- knowledge and skills related to growing you own are gradually lost.

In other words, the people are invaded by forces that they know little about, and don't understand. Their self-reliance is undermined. They become easy meat for even more invaders, both biological and social. They are exploited and manipulated. Of course, you (the reader) are not subjected to this kind of invasion. You are not influenced by corrupt politicians. You are not influenced by advertising of food and drink. You would not allow your environment to be altered by corporates. You would not allow your common sense about food to be undermined. You are immune from these invaders. Good luck to you.

The evolution of human food

The evolution of human food is continuing apace. If we look back 10,000 years, or thereabouts, human hunter-gatherers faced natural hazards, such as poisonous plants, predators, and parasites. They differed from most other animals in having weapons, fire, and knowledge that could be passed from one generation to another. A few such peoples still exist today, but most present-day humans had ancestors that used their knowledge to domesticate plants and animals, and their diet began to change.

The change in the human lifestyle from a nomadic to a settled existence had profound, irreversible consequences. Hunting and gathering did not vanish, but it became less productive, because people could not move far from their settlements, and follow the game and the seasons as they used to. Their diet now contained a much higher proportion of carbohydrates, and they became victims of various digestive problems, including milk intolerance and gluten intolerance[8]. Their lives were shorter, but their reproduction higher. Consequently, the rate of growth of human populations increased.

The other big change was the loss of self-sufficiency for an increasing proportion of the population. If you had no land or animals, you had to buy your food. This change has happened gradually, over centuries, but it is inexorable. More and more people have to buy their food with money, and to do this theyhave to earn money. In many modern cities poor people have a poor diet, because they cannot afford proper food.

Deterioration of the human diet is on the increase for three main reasons. Firstly, the more people that rely on food providers, instead of growing their own, the more power the food providers have. You can buy good food, at a price. If you can't afford the price then you have to eat inferior food, unless you have the knowledge and initiative to circumvent this problem. Indeed, there is no end of advice available, in doctor's surgeries, on TV, etc., but there is still a relentless dumbing down of the urban diet.

Secondly, food providers have a long history of being deceitful and manipulative. In the early days it was food adulteration that was the main problem (see Chapter 5). When this was stamped on by governments, the food providers began to over-advertise their products, claiming all sorts of spurious benefits, often making pseudo-scientific claims (see Chapter 6). As these practices have become more regulated by governments, food providers have attempted to manipulate the public, by making their products addictive (e.g. alcopops), by attempting to make their products fashionable (see Chapter 9), and by shouldering out the opposition (mainly small businesses) by altering the environment (see how the British high street has changed over the past 30 years). In many western cities it is difficult to buy food, except from supermarkets.

Thirdly, there have been global changes in the environment, and in the production of food. Vast swathes of land have been cleared of their natural vegetation, not to grow food, but to grow money in one way or another. Some of this land is used to grow cash crops, some is flooded, some is used as building land, and some is mined or deforested for timber. Some of this land exploitation is done by governments (in our name), but the major damage is done by corporates. The result is that, having survived decades of declining food quality, we are now about to face a period of declining food quantity. Food prices are rising all over the world[9].

Do we have the power to reverse this trend? Colin Tudge, in his book *Feeding People is Easy* says yes.

"The message of this book is as positive as anyone could hope for: the future could still be glorious. We have just to do things differently."[10]

How can we do things differently? Do we have the means to change things? Our power to change things can be measured in terms of Purchasing Power Parity (PPP) calculations. PPP takes into account the relative cost of living and the inflation rates of people in different countries. Now, in some countries, governments are accountable to their public, because people can vote. In other countries the governments are not accountable, because people cannot vote, or their votes have no effect. At the present time, roughly two-thirds of the PPP in the world comes from people who have a vote that is effective. The other third comes from people who have no effective vote, plus the corporates. So it would appear that despite opposition from corporates, the majority of people in the world, if so minded, could accomplish what needs to be accomplished. The problem is the 'if so minded requirement'. If we allow our minds to be invaded, and manipulated, then we will probably not be so minded, at least, not sufficiently.

The problem is that parasites are hard to eradicate, particularly those that invade the mind. Just like the rats that lose their wariness of cats, when the parasitic protozoan invades its brain (see Chapter 6), so we lose our wariness of food when advertisements invade our brain. The parasite manipulates the rat to do something that it would not otherwise do. Advertisers persuade us to buy something that we would not otherwise buy. Of course, if we did not know about it we could not buy it. But often we do know, but we don't care. Moreover, the corporates will say that they are providing us with what we want. They are offering to improve our lives. We have a choice. But what rat would not choose to lose its fear of cats? What rat would not choose an easier (but shorter) life? Maybe the rat does know that cats are dangerous, but does not care. Maybe you do know that certain eating habits are dangerous, but do not care. Maybe you do prefer the easier (but shorter) life.

Glossary: A guide to the meaning of technical terms as used in this book

Accessibility The ability of an agent to obtain available resources by virtue of its skills, tools, etc. (technically – a limit on the rate at which it can obtain resources).

Addiction Physiological and/or psychological dependency on practices that are not necessary for good health, and may be harmful. Addictive practices include ingestion of harmful substances, smoking and gambling.

Anticipatory drinking Drinking in advance of physiological dehydration, in circumstances that are likely to promote dehydration

Anorexia nervosa A psychiatric illness involving an eating disorder characterized by extremely low body weight and body image distortion with an obsessive fear of gaining weight.

Artificial selection Evolution of traits through selective breeding, as in the domestication of animals by humans.

Availability Ability of the environment to provide a resource.

Broadcast information is a measure of the information obtained from a signal by an observer, particularly with respect to the sender's identity and behavior.

Budget The limit to expenditure of energy or time that constrains behavior.

Co-adaptation Mutual adaptation of separate structures, or of aspects of behavior, that are designed by natural selection specifically for interaction with each other. Co-adaptation may occur between parts of a single organism, or between organisms, or between species.

Common currency A quantity common to both items of a choice situation. Thus in deciding between chalk and cheese, such a quantity might be weight, hardness, etc.

Communication Transmission of information from one individual to another which is designed to influence the behaviour of the recipient.

Community An association of plants and animals living together in a particular *habitat.*

Competitive exclusion principle Two species with identical niches cannot live together in the same place at the same time when resources are limited. The corollary is that, if two species coexist, there must be ecological differences between them.

Conditioning A process involved in learning, in which an animal forms an association between a previously significant stimulus and a previously neutral stimulus or response. Thus an association may be formed between a significant stimulus, such as the sight of food, and a neutral stimulus, such as a sound, or the time of day. Initially, the animal responds only to the food, (e.g. by salivating) but if the same situation occurs on a number of occasions, then the animal comes to associate the sound (or the time) with the food, and will eventually salivate even in the absence of food. The response of the animal is said to have become conditioned to the sound (or time).

Cue strength   The combination of external stimuli that give rise to a particular activity..

Cultural transmission   Behavior that has been passed from one generation to another by non-genetic means.

Death by eating   Death caused by anything associated with eating, such as the food ingested, injury or disease.

Deceit   Communication that benefits the sender at the expense of the receiver

Demand   An economic concept that expresses the relationship between the price and consumption of a commodity. When applied to animals, the price is usually expressed in terms of energy, and consumption in terms of time spent on the behaviour in question.

Discrimination   Making different responses to different stimuli.

Eavesdropping   A form of manipulation in which benefit is gained by intercepting the communication of others. For example, a rival may benefit from a male's message to a female, because it leads him to the female.

Ecological succession   Changes in the composition or structure of an ecological community as a result of a major disturbance, such as fire, storms or human intervention.

Essential nutrient   A nutrient required for normal body functioning that cannot be synthesized in amounts adequate for good health. They must be obtained from a dietary source. Different species may have different essential nutrients. For example, most mammals synthesize their own ascorbic acid and it is therefore not considered an essential nutrient for such species. It is, however, an essential nutrient for human beings, who require external sources of ascorbic acid (known as Vitamin C in the context of nutrition).

Evolution   Origination of species characteristics by development from earlier forms. Now acknowledged to be due to the process of natural selection.

Evolutionary discordance   A discordance between the present situation and the situation to which the organism achieved evolutionary adaptation.

Evolutionary strategy   A passive result of natural selection that gives the appearance of a ploy employed by genes to increase their numbers at the expense of other genes. An evolutionary strategy is not a strategy in the cognitive sense, but a theoretical tool employed in evolutionary theory.

Evolutionary theory   A body of theory derived from the principle of natural selection.

Extended phenotype   An aspect of the phenotype that is outside the body, but nevertheless a result of genetic influence, such as the type of nest typically made by a species of bird.

Extracellular fluid   The fluid surrounding cells.

Fetal origins   The hypothesis that the origin of certain traits, such as body weight or food preferences, originate from influences in the womb.

Fitness   A biological and mathematical concept that indicates the ability of genetic material to perpetuate itself during the course of evolution.

Food-aversion conditioning   Learning (by conditioning) to avoid food that has deleterious physiological consequences when eaten.

Food chain   An aspect of community life, in which each species occupies a particular niche that includes food from species further down the chain. Thus each species is an essential component of the community, necessary for the survival of the species that eats it.

Genotype   The genetic makeup of an individual.

Globalization   An ongoing process of trans-national dissemination of language and ideas, especially in relation to the spread of technology and of economic developments. Direct investment by corporations into foreign countries is an aspect of globalization that has particular effects upon food supply..

Habitat   The natural home of an animal or plant: the external environment to which it has become adapted during the course of evolution. Habitats are usually described in terms of salient physical and chemical features of the environment, and largely determined by climate.

Homo neanderthalis   A species of human that used to inhabit the earth.

Homo sapiens   The species of humans that currently inhabits the earth.

Incentive motivation   An aspect of motivation that derives from external stimulation. Thus the sight of food (and its apparent palatability) contribute to appetite.

Insulin   A hormone that causes cells in the liver, muscle and fat tissue to take up glucose from the blood, and store it as glycogen in the liver and muscle.

Labor supply curve   A curve joining points on a graph of wage rate against hours. It shows how a change in wage rate affects the number of hours (per day) that a worker is willing to work.

Learning   An irreversible change in response to particular stimuli, as opposed to the reversible changes that result from changes in motivation. Learning is characterised by changes in memory, unlike other irreversible changes, such as those due to maturation and injury.

Manipulation   An aspect of communication which can occur when the receiver obtains information about the signaller, against the interests of the signaller. For example the courtship display of a male may attract rivals.

Maturation   An irreversible part of individual development. Maturation does not depend upon experience, and is thus distinct from learning and injury. For example, pigeons have been reared in confined conditions so that they cannot move their wings. When released, these birds could fly just as well as other birds of the same age. Although young birds can be seen flapping their wings whilst remaining on the ground, this apparent practice has no effect upon the development of flight.

Mental state   A state of mind that is about something, or refers to something.

Message   A signal as decoded by the receiver.

Metabolism   The set of chemical reactions that occur in living organisms to maintain life.

Mimicry   The resemblance of animal (the mimic) to another (the model) such that the two are confused.

Motivation   A reversible aspect of the animal's state that plays a causal role in behaviour. For example, hunger and sexual arousal are causes of behaviour that are reversible, whereas changes due to learning, maturation and injury, are not reversible

Motivational priorities   Those activities that an animal is most motivated to perform at a particular time, even though it can perform only on of them at a time.

.

Motivational state   The state that results from the combination of external and internal motivational changes.   Changes in motivational state may be exogenous, emanating from changes in the outside world that are perceived by the animal, or endogenous and stemming from the animal's own behaviour. In both cases the motivational state of the animal is changed as a result of stimuli impinging on the animal's internal and external sensory systems.

Multidimensional motivational state   A motivational state that involves more than one type of behavior. Thus hunger may require a variety of foods to achieve satiation.

Natural selection   The principle underlying the evolution of plants and animals.

Neophobia   Aversion to novel situations.

Niche The role played by a species in the community, in terms of its relationship both to other organisms and to the physical environment.

Niche overlap The situation that pertains when animals of different species use the same resources or have certain preference or tolerance ranges in common.

One-dimensional motivation Motivation that implicates only one type of behavior. Thus thirst can be slaked by a certain amount of water, but not by different types of water.

Opportunity A favourable set of circumstances that is accessible to the animal. At the species level, this generally applies to the exploitation of new habitats. At the individual level, opportunism often takes the form of rapid exploitation of new sources of food.

Optimal A criterion in relation to which it is possible to determine which of a set of alternatives is the best.

Phenotype The bodily expression of genetic influence, and of characteristics acquired by learning and culture.

Pheromones Chemicals that are released into the environment by an organism, and cause a specific behavioural or physiological response to a receiving organism (usually of the same species).

Receiver The recipient of a signal that is capable of decoding it into a message.

Reproductive success An aspect of Darwinian fitness that is a measure of an animal's surviving offspring.

Ritualization A process of evolution by which behaviour patterns become modified to perform a communication function.

Satiety The state of zero hunger, signalled by a combination of messages from the body.

Sexual selection A form of natural selection which depends upon the advantage that certain individuals have over others of the same sex and species, solely in respect of reproduction.

Sign Construct by which an organism affects the state of another.

Signal The physical embodiment of a message.

Specific hunger A type of hunger that is satisfied by specific dietary requirements, such as vitamins and minerals.

Source The origin of an encoded signal.

Stratified redistribution Distribution of pooled food on the basis of some social criterion.

Supernormal stimulus A stimulus that surpasses a natural stimulus in its effectiveness. Animals often have an open-ended preference for stimulus properties, such as size, sweetness, of color, that is limited in the natural world. When presented with an artificial supernormal stimulus they prefer it to the natural one.

Survival value The survival of a trait within a population in the face of the selective pressures inherent in the environment

Symbiosis   The living together of organisms of different species to their mutual benefit. It occurs amongst plants, between plants and animals, and among animal species. It can occur as a relationship between individuals, between individuals and societies, and even between societies.

Territory   A patch of ground that is defended against members of the same species. The defended area may function as a protected source of food, or mating partners, or places to raise the young.

Tolerance   The ability to tolerate extreme values of environmental factors, such as temperature, humidity, etc.

Trade-off   Balancing of priorities in such a way as to obtain the best (optimal) overall result.

Transmitted   information is a measure in terms of the increase in the predictability of the receiver's behavior following activity by the sender.

Warning   An aspect of communication that functions to ward off predators, or to alert others to danger, The warning may take the form of a specific visual, auditory, or olfactory alarm signal.

Utility   A term used in to denote the quantity maximised by the individual animal, or person, in the process of rational decision-making. As in human terms, utility is a notional measure of the psychological value of the consequences of an action (e.g. buying goods).

END NOTES

Chapter 1

1. Jared Diamond, *Guns,Germs and Steel.* Vintage 1998, pp 39-52

2. Our phenotype is the bodily expression of genetic influence, and of characteristics acquired by learning and culture. Our extended phenotype includes our tools and artefacts, our clothes and our food. It used to be thought that characteristics acquired by learning could not influence the genes that are passed on to the next generation. This view has recently been questioned (Laland, K.N, , Odling-Smee, J. and Myles, S. 2010 How culture shaped the human genome: Bringing genetics and the human sciences together. *Nature Reviews Genetics* 11: 137-148, 2010.)

3. Scientists distinguish between two types of primary thirst. *Extracellular thirst* results from a reduction in the volume of fluid found between cells. This type of thirst can be referred to as volumetric thirst or hypovolemia. Volumetric thirst can be caused by a number of things including bleeding, vomiting, diarrhea, sweating, and alcohol consumption. It arises when the volume of blood plasma, i.e. intravascular fluid, decreases. As intravascular fluid decreases, blood pressure is reduced and the body attempts to compensate by moving fluid from other cellular compartments into the blood vessels. Fluid is transferred from all of the fluid compartments in the body. Pure volumetric thirst is caused by the loss of blood and because sodium is also lost from the plasma, the body's need for salt proportionately increases.

Intracellular thirst is triggered when fluid is drawn out if cells due to an increase in the concentration of salts and minerals outside of the cell. It can be caused by salty snacks etc. Intracellular thirst is sometimes referred to as osmometric thirst. It is produced by an increase in the osmotic pressure of the interstitial fluid relative to the intracellular fluid, and is also known as cellular dehydration. Osmometric thirst also occurs when the solute concentration of the interstitial fluid increases. This increase in solute draws water outside of the cell, causing the cell to shrink. Similar to osmosis, osmometric thirst can be correlated with the movement of water through a semipermeable membrane from an area with a low concentration of solute to an area with a high solute concentration. Osmoreceptors in the brain are neurons that adjust their firing rate in relation to their level of hydration or lack thereof. Accordingly, as fluid surrounding the cell becomes more concentrated, water flow out of the cell causing it to shrink. The change in the size of the cell causes the osmoreceptors to change their rate of firing, sending signals to various areas of the brain.

As a person eats a salty meal, he or she sustains pure osmometric thirst. Salt is absorbed from the digestive system into the blood plasma causing it to become hypertonic. A hypertonic solution refers to a solution that contains enough solute to draw water out of the cell through the process of osmosis. It is this condition that draws water from the interstitial fluid causing it too to become hypertonic. Because the level of solute is greater outside the cell, water leaves the cell and the cell shrinks, causing a change in the firing of the osmoreceptors thereby initiating osmometric thirst. Drinking water is the best way to reduce this thirst.

In humans, dehydration can be caused by a wide range of diseases and states that impair the water balance of the body.

These include stress-related causes, survival situations, blood loss due to physical trauma, diarrhoea, vomiting, burns and infectious diseases such as cholera, gastroenteritis and yellow fever.

Symptoms may include headaches similar to those experienced during a hangover, decreased blood pressure, and dizziness or fainting when standing up. Untreated dehydration generally results in delirium, unconsciousness, and in extreme cases death.

Dehydration symptoms generally become noticeable after 2% of one's normal water volume has been lost. Initially, one experiences thirst and discomfort, possibly along with loss of appetite and dry skin. This can be followed by constipation. Symptoms of mild dehydration include thirst, decreased urine volume, abnormally dark urine, and unexplained tiredness, lack of tears when crying, headache, and dry mouth.

The best treatment for minor dehydration is drinking water and stopping fluid loss. Water is preferable to sport drinks and other commercially-sold rehydration fluids, as the balance of electrolytes they provide may not match the replacement requirements of the individual

4. These experiments were done by my student Phil Budgell (1970) Modulation of drinking by ambient temperature changes. *Animal behaviour*, 18, 753-757, and (1970) The effect of changes in ambient temperature on water intake and evaporative water loss. *Psychonomic Science,* 20, 275-276. Despite their obvious relevance to the study of homeostasis, the message has still not penetrated into medical and school textbooks.

5. The cells of the nervous system can absorb glucose in the absence of insulin, but other cells require insulin to transport glucose across the cell wall. During times of glucose shortage, as in fasting, the level of insulin in the blood falls to such an extent that the blood glucose is effectively available only to the cells of the nervous system. The other cells have to depend upon metabolism of fatty acids to obtain energy. During fasting, glucose is derived from body reserves, which include glycogen stored in liver and muscle, fat stored in various parts of the body, and in the last resort, the protein of muscle and other tissues. Fat is broken down to glycerol and fatty acids, and the glycerol is converted to glucose in the liver. Protein is broken down into amino acids, and these are metabolised by the liver to produce some glucose.

6. Salt deficiency may be caused by lack of salt in the diet, hormonal imbalances (involving vasopressin or ADH), loss of blood, and excess sweating. Athletes (e.g. marathon runners) may lose salt in their sweat, and if they respond to this by drinking only water, they may develop water intoxication (causing an osmotic shift of water from the plasma into the brain cells).

7. An interesting account of the importance of salt can be found in Reay Tannahill's *Food in History* , Paladin, 1975, pp 180-188.

8. Excess salt in the diet can lead to hypertension, osteoporosis and stomach cancer.

9. Thiamine (vitamin B_1) plays an important role in carbohydrate and fat metabolism. It is essential for normal growth and development, and for the proper functioning of the heart and the digestive and nervous systems. Thiamine

deficiency can lead to degeneration of the nervous system, and diseases such as beriberi.

10. Rozin, P. Specific aversions as a component of specific hungers. *J. comp. Physiol, Psychol.,*64, 237-242. 1967. Rozin, P. and Kalat, J. Specific hungers and poison avoidance as adaptive specializations of learning. *Psychol. Rev.*78, 459-486, 1971.

11 A persistent tendency to put things in the mouth, or to attempt to eat inedible substances (e.g., coal, soil, chalk, paper etc.), is called pica. In order for these actions to be considered pica, they must persist for more than one month, at an age where eating such objects is considered to be inappropriate. Pica is seen in all ages, but particularly in pregnant women and small children. Pica has a number of causes, dietary deficiency being one of them.

12. An essential nutrient is a nutrient required for normal body functioning that either cannot be synthesized in amounts adequate for good health. They must be obtained from a dietary source. Different species may have different essential nutrients. For example, most mammals synthesize their own ascorbic acid and it is therefore not considered an essential nutrient for such species. It is, however, an essential nutrient for human beings, who require external sources of ascorbic acid (known as Vitamin C in the context of nutrition). Many essential nutrients are toxic in large doses. Some can be taken in amounts larger than required in a typical diet, with no apparent ill effects

.Essential nutrients include the Linolenic fatty acids; the essential amino acids (Isoleucine, Lysine, Leucine, Methionine,

Phenylalanine, Threonine and Valine), plus some essential amino acids necessary for human children but not adults (Histamine, Arginine); the dietary minarels (Calcium, Chloride, Cobalt, Copper, Iodine, Iron, Magnesium, Molybdenum, Nickel, Phosphorus, Potassium, Selenium, Sodium, Sulfur, and Zinc); and the vitamins .(Biotin, Choline, Folate, Niacin, Pantothenic, Riboflavin, Thiamine, Vitamin A (retinol), Vitamin B6, Vitamin B12, Vitamin C (ascorbic acid), Vitamin D, Vitamin E and Vitamin K)

13. A trade-off is a balancing of priorities, either as an aspect of design (e.g. in designing a bone (by natural selection) there is a trade-off between strength and weight. In designing a bicycle (by deliberation) there is also a trade- off between strength and weight) or of mechanism (e.g. motivational priorities).Foraging efficiency is usually a matter of trade-off among competing motivational priorities. These may include energy gained versus energy spent, energy gained versus risk if predation, and energy gained versus losses to rivals.

14 Giles Milton (1999) *Nathaniel's Nutmeg* . Hodder and Staughton, London.

15. See, for example Bergeron, R. Badnell-Waters, A.J., Lambton, S. and Mason, G. Stereotypic oral behaviour in captive ungulates: foraging, diet, and gastrointestinal function. G. In Mason, G. and Rushen, J. (eds) *Stereotypic Animal Behaviour. Fundamentals and applications to welfare.* 2nd Edition, CABI, Wallingford, Oxfordshire, 2006.

16. Drent, R.H. (1970) Functional aspects of incubation in the herring gull. In Baerends, G.P. and Drent, R.H. (eds) The herring gull and its egg, Part 1. *Behaviour,* 82, 1-132. Sibly, R.M. and McCleery, R.H. (1985) Optimal decision rules for gulls. *Animal Behaviour*, 33. 449-65.

17. Lettvin,J.W, *et al.* (1959) What the frogs eye tells the frogs brain. *Proc. I.R.E.* ,47, 1940-1951.

18   For an account of this voyage see Giles Milton (1999) *Nathaniel's Nutmeg* . Hodder and Staughton, London. Pp 42-52.

Chapter 2

1. For a fuller account see Owen, J. (1980) *Feeding Strategy,* Oxford University Press

2. For an engaging account of the complex relationships between figs and wasps, see Tudge, C. (2005) *The secret life of trees.* Penguin books, London.

3. For a more detailed account of plant-animal interactions, see Herrera, C.M. and Pellmyr, O.(eds) (2002) *Plant-animal interactions: an evolutionary approach.* Blackwell Science, Oxford.

4. Dowd, P. 1991. Symbiont-mediated detoxification in insect herbivores. Pages 411 – 440 in P. Barbosa, V. A. Krischik, and C. Jones,(eds.) *Microbial mediation of plant – herbivore interactions.* Wiley & Sons, Inc., New York, USA. Krokene, P., and H. Solheim. 1998. Pathogenicity of four blue-stain fungi associated with aggressive and nonaggressive bark beetles. *Phytopathology* 88:39 – 44. Whitney, H. S. 1982. Relationships between bark beetles and symbiotic organisms. Pages 183 – 211 in J. B. Mitton and K. B. Sturgeon, editors. *Bark beetles in North American conifers.* University of Texas Press, Austin, Texas, USA.. Nebeker, T. E., J. D. Hodges, and C. A. Blanche. 1993. Host response to bark beetle and pathogen colonization. Pages 157 – 173 in T. Schowalter,

editor. *Beetle-pathogen interactions in conifer forests.* Academic Press, New York, USA.. Sagers, C. L. 1992. Manipulation of host plant quality: Herbivores keep leaves in the dark. *Functional Ecology* 6:741 – 743.

5. Neanderthals (*Homo neanderthalensis*) had many adaptations to a cold climate, such as large braincase, short, robust build, and rather large noses (to heat incoming air). Compared to modern humans, Neanderthals were similar in height but with more robust bodies, and had distinct morphological features. Evidence suggests that they were much stronger than modern humans. A 2007 study confirmed that some Neanderthals had red hair and pale skin colour, as would be expected of people living in areas of reduced sunlight. The mutation in the gene responsible arose independently of the mutation that causes a similar pigmentation pattern in modern humans.

6. Robson, A.J., and Kaplan, H. "The Evolution of Human Life Expectancy and Intelligence in Hunter-Gatherer Economies." *American Economic Review,* 93 (1): 150-169. 2003.

7. The Neanderthals had fire, tools and weapons, including spears with long wooden shafts and stone heads, and stone knives and scrapers. They were great meat eaters, and hunted mammoths. These animals are evolved from Asian elephants, and they were probably susceptible to elephant pox, a virus closely related to cow pox, and small pox. It is known that elephants can infect people with the virus, but most elephant handlers acquire immunity. See Kubin, G.; Kolb, O.; Gerstl, F. Wien, Characterization of a pox virus strain isolated from an elephant. *Tierarztl Monatsschr*, 62, 271-276. 1975; and Kurth A, Wibbelt G, Gerber H-P, Petschaelis A, Pauli G, Nitsche A. Rat-to-elephant-to-human transmission of cowpox virus

Emerg Infect Dis 2008. T.D. Berger and E. Trinkaus, Patterns of trauma among Neandetthals.*Journal of Archaeological Science* **22**: 841 - 852. 1995.; Richards, M.P., Pettitt, P.B. , Trinkaus, E., Smith F.H., Paunovi M. and Karavani, I Neanderthal diet at Vindija and Neanderthal predation: The evidence from stable isotopes Published online on June 13, 2000, 10.1073/pnas.120178997 *PNAS* , June 20, 2000,. 97. 13 | 7663-7666 , 2000; Conard,N.J. and Niven, L. The Paleolithic finds from Bollschweil and the question of Neanderthal mammoth hunting in the Black Forest. *The World of Elephants - International Congress, Rome 2001*; Laleuza-Fox, C., Holger R. *et al*. A Melanocortin 1 Receptor Allele Suggests Varying Pigmentation Among Neanderthals". *Science*, 10,1126, 2007

Chapter 3

1. Heinrich, B. *Bumblebee economics*. Harvard University Press, Cambridge, MA. (1979).

2. The field work is due to Richard Lee *The !Kung San.* Cambridge University Press. (1979) and Lee, R. and DeVore, I (eds) *Man The Hunter*, Chicago, Aldine. (1968). For the wage rate analysis of the San, see McFarland, D. *Problems of Animal Behaviour,* Longman, Harlow UK (1989) pp 59-86.

3. For a fuller account see Sahlins, M, *Stone Age Economics*. Tavistock Publications, London, (1974) pp. 1-39. For the origins of plant domestication see Jared Diamond, *Guns, Germs and Steel.* Vintage 1998, pp 176-191.

4. Useful references are Kagel J.H., Battalio R.C., Rachlin H. and Green L., H. and Demand Curves for Animal Consumers,1981, *Quarterly Journal of Economics*, vol. 96, 1-15. Stephens, D. W. and Krebs, J.R. Foraging theory, Princeton

University Press, 1986. Houston, AI. (1997) Demand curves and welfare. *Animal Behaviour*, 53, 983 - 990

5. For further details, see Bell, W.J. (1991) *Searching Behaviour: The behavioural ecology of finding resources.* Chapman and Hall, pp 217-224.

6. For the fieldwork on redshank see Goss-Custard, J.D. Optimal foraging and size selection of worms by redshank

(*Tringa totanus*). *Animal Behaviour*, 25, 10-29, 1977. and Goss-Custard, J.D. Predator responses and prey mortality in the redshank *Tringa Totanus* (L) and a preferred prey *Corophium volutator* (Pallas). *Journal of Animal Ecology*, 46, 21-36, 1977. For an economic analysis see McFarland, D. *Animal Behaviour* (third edition) Longman, Harlow, UK. (1999) pp 435-441.

7. For a review, see Barnard, C. The evolution of food-scrounging strategies within and between species. In *Producers and Scroungers, Strategies of Exploitation and Parasitism.* (ed) C.J. Barnard, Croom Held, London, pp 95-126.,1984.

8. Bertram, B. C.R. Vigilance and group size in ostriches. *Animal Behaviour*, 28, 278-286.1980. Davis, J.M. Socially induced flight reaction in pigeons. *Animal Behaviour*, 23, 597-601, 1975.

9. Alarm signals are responses to signs of danger that act as a warning to other animals. Some alarm responses may attract the attention of the predator, and thus endanger the animal giving the alarm. This poses a problem for evolutionary theory, because those individuals that readily give alarm signals would seem to be disadvantaging themselves. There would seem to be an advantage in being a 'cheat' and relying on others to watch out for predators and give the alarm. So after some generations

cheaters should become more numerous, and the group as a whole more prone to predation. This is called the problem of altruism (self-destructive behaviour performed for the benefit of others).In many species, however, those warned by the alarm-giver are kin. That is, they carry many of the same genes as the alarm-giver, and these genes are likely to include the 'altruistic' gene. So from the genetic point of view, the alarm-giver may be endangering itself as an individual, but its behaviour is helping to ensure that the alarm-giving genes are maintained within the population. This process is usually called kin selection.

10. Mackenzie, D. *Goat Husbandry*, (3rd edition) Faber and Faber, London, p. 109. 1970.

11. Ethnobiology is a relatively new anthropological discipline that focuses on how indigenous people learn, name, use, and organize knowledge about the plants and animals around them. It is sometimes called folk biology. Ethnobiology is the study of the past and present interrelationships between human cultures and the plants, animals, and other organisms in their environment, including relationships with ecosystems as a whole. It is an interdisciplinary subject which draws on knowledge from many different fields of knowledge such as linguistics, anthropology, biology, chemistry. For further information, see Ellen, R. (ed) *Ethnobiology and the Science of Human Kind*, Blackwell, Oxford. 2006.

12. Jared Diamond, *Guns,Germs and Steel*. Vintage 1998, pp 39-52

13. *Narcissus* bulbs can be easily mistaken for onions. Ingestion of the bulbs (even when cooked) produces abdominal pains, vomiting, nausea, light-headedness, shivering, and sometimes diarrhoea, and convulsions which

may be fatal. The toxic principles are Phenanthridine alkaloids such as lycorine, also calciumoxalate crystals. Cooper, M. R., Johnson, A. W. 1984. *Poisonous plants in Britain and their effects on animals and man.* Her Majesty's Stationery Office, London, England. 305 pp Gonçalo, S., Freitas, J. D., Sousa, I. 1987. Contact dermatitis and respiratory symptoms and *Narcissus pseudonarcissus. Contact Dermatitis,* 16: 115-116.Litovitz, T. L., Fahey, B. A. 1982. Please don't eat the daffodils. *N. Eng. J. Med.,* 306: 547

Chapter 4

1. The dates of the major human migrations are constantly under revision. For a recent update on the chronology of human routes out of Africa, based upon DNA studies, see Jones, D. Going global. *New Scientist,* 196, 36-41, 27 Oct 2007. See also Cavalli-Sforza, L.L. *Genes, peoples and Languages.* Penguin Books, 2000.

2. Childe, G. *What Happened in History, London* (p43) 1942. Childe, G. *The Prehistory of European Society, London,* (p35) 1958.

3. Good accounts of domestication are given by Ucko, P.J. & Dimbleby, G.W. (eds) *The domestication and exploitation of plants and animals.* Gerald Duckworth & co., London, 1969, and Jared Diamond, *Guns, Germs and Steel.* Vintage 1998. The dates given here are from various later works. Good accounts of the domestication of plants can be found in Ucko, P.J. & Dimbleby, G.W. (eds) *The domestication and exploitation of plants and animals.* Gerald Duckworth & co., London, 1969, and Jared Diamond, *Guns, Germs and Steel.* Vintage 1998. and especially in Zohary, D and Hopf, M (1993) *Domestication of plants in the Old World .* Clarendon Press, Oxford.

4. Archaeologist J.R.Harlan discovered this by trying it himself in the mid 1960.(see Tannahill, R. *Food in History*. Paladin 1975, p36-54).

5. Jared Diamond, *Guns,Germs and Steel*. Vintage 1998, pp 362-3

6. It may well be that the cultivation of rice in China and Japan preceded the cultivation of cereals in the rest of the world. For a review, see Yoshinori, Y. (ed) *The Origins of Pottery and*

Agriculture. Yoshinori Yasuda (Ed). New Delhi: Roli Books Pvt. Ltd. , 2002,

7. In 1871 Darwin published *The Descent of Man,* in which he considered the subject of sexual selection, to which he had referred to in his *Origin of Species* (1859). According to Darwin, sexual selection depends upon the advantage which certain individuals have over others of the same sex and species solely in respect of reproduction. Darwin reasoned that females make a definite choice of sexual partner and that males have acquired particular adornments and courtship behaviour 'not from being fitted to survive in the struggle for existence but from having gained an advantage over other males, and from having transmitted this advantage to their male offspring alone'. Sexual selection is a complex subject that biologists have been arguing about for decades. For a good account, see Cronin, H. *The Ant and the Peacock*. Cambridge University Press, 1991.

8. The origin of the founder mutation probably occurred in the Black Sea region, where the great agriculture migration to the northern part of Europe took place in the Neolithic periods about 6–10,000 years ago The high frequency of blue-eyed individuals in the Scandinavia and Baltic areas indicates a

positive selection for this phenotype. Several theories have been suggested to explain the evolutionary selection for such pigmentation traits, including the effects of ultra-violet light, vitamin D and sexual selection. For further information see Eiberg, H., Troelsen, J., Nielsen, M., Mikkelsen, A., Mengel-From, J., Kjaer, K. W. and Hansen, L. Blue eye color in humans may be caused by a perfectly associated founder mutation in a regulatory element located within the HERC2 gene inhibiting OCA2 expression. *Human Genetics* (2008) 123:177–187.

9.   The enzyme lactase, also called beta-D-galactosidase, is synthesized if at least one of the two genes for it is present. Only when both gene expressions are affected is lactase enzyme synthesis reduced, which in turn reduces lactose digestion.. Lactase persistence, allowing lactose digestion to proceed, is the dominant allele. Physiological lactose intolerance, therefore, is an autosomal recessive trait. Lactose is a water soluble molecule, and when milk is separated into curds and whey, it is found in the water whey and not in the fatty curds. So, because the butter making separates milk's water components from the fat components, lactose will not be present in the butter. Cheeses vary in their lactose content, depending upon the method of cheese making, but traditionally made yogurt contains lactase enzyme produced by the bacterial cultures used to make the yogurt. So lactose intolerant people can usually digest yogurt.

The prevalence of lactose intolerance varies markedly amongst modern human populations, and is related to their ancestral genetic makeup. It is very low in people with Scandinavian ancestors, including the Dutch, British and White North Americans, and very high in people of Mongolian descent, including the Chinese, other Asians and Native Americans.

10. The evidence comes from calculations about the rate of evolutionary change. The rate of change of gene frequency induced by natural selection depends upon the relative fitness of the various genotypes. The difference between two genotypes of differing fitness can be used to calculate the coefficient of selection against the inferior genotype, and this can be used to calculate the rate of change of phenotype within a population. In the case of the genotypes involved in lactose tolerance in human adults such calculations indicate that the relevant genetic changes are entirely possible within the 10,000 year time frame. (For a brief account see McFarland, D. Animal Behaviour (3rd edition) Longman, 1999, pp 59-61. For substantive accounts see Bodmer, W.F. and Cavalli-Sforza, L. *Genetics, Evolution and Man.* W.H.Freeman, San Francisco, 1976.). Flatz G (1987). Genetics of lactose digestion in humans. *Adv. Hum. Genet.* 16: 1–77. Swallow D.M .(2000). Genetics of lactase persistence and lactose intolerance. *Ann. Rev. Genet.* 37: 197–219.

11. Evolution acting through natural selection represents an ongoing interaction between a species' genome (genetic makeup) and its environment over the course of many generations. Genetic traits may be positively or negatively selected relative to their concordance or discordance with environmental selective pressures. When the environment remains relatively constant, stabilizing selection tends to maintain genetic traits that represent the optimal average for a population. When environmental conditions permanently change, evolutionary discordance arises between a species' genome and its environment, and stabilizing selection is replaced by directional selection, which changes the average population genome. Initially, when permanent environmental changes occur in a population, individuals bearing the previous average status quo genome experience evolutionary discordance. In the affected genotype, this evolutionary

186

discordance may manifest itself phenotypically as disease, increased morbidity and mortality, and reduced reproductive success.

Key references are: Voegtlin, W. L. *The stone age diet: Based on in-depth studies of human ecology and the diet of man.* Vantage Press 1975. Eaton SB, Konner M (1985) Paleolithic nutrition: a consideration of its nature and current implications. *New England Journal of Medicine,* vol. 312, pp. 283-289. Eaton SB, Konner M, Shostak M (1988) Stone agers in the fast lane: chronic degenerative diseases in evolutionary perspective. *American Journal of Medicine,* vol. 84, pp. 739-749. Eaton, S. Boyd; Eaton, Stanley B. III; Konner MJ; Shostak M (1996) An evolutionary perspective enhances understanding of human nutritional requirements. *Journal of Nutrition,* vol. 126 (1996), pp. 1732-1740. Eaton SB, Eaton SBI, Konner MJ. Paleolithic nutrition revisited. In: Trevathan WR, Smith EO, McKenna JJ, eds. *Evolutionary Medicine.* New York, NY: Oxford University Press; 1999: 313.

12. Good references are Nentwig,W. Human Environmental Impact in the Paleolithic and Neolithic.*Handbook of Paleoanthropology,* Henke,W. and Tattersall. I. (eds) Springer, Berlin, 2007, and Williams, M. *Deforesting the Earth: From Prehistory to Global Crisis.* University of Chicago Press, 2002.

13. For further information see Gavrilov, L.A. and Gavrilova, N.S. *The biology of life span: A quantitative approach.* Harwood Academic Publisher, New York, 1991.

14. There were differences in the timing of the Neolithic Revolution across regions, and these generated significant variations in the genetic composition of the contemporary human populations. Combined with skeletal evidence, a picture

emerges of the lifestyle of Neolithic farmers. For further details see Steckel (2004), Mokyr (1998), (e.g., Cohen, M. N. *Health and the Rise of Civilization,* Yale University Press: New Haven, 1989. For the effect of the Neolithic Revolution on the exposure and the vulnerability of humans to environmental hazards such as infectious diseases see Diamond, J., *Guns,Germs and Steel.* Vintage 1998 and, Weisdorf, J.L. *From Foraging to Farming: Explaining the Neolithic Revolution,* Discussion papers 03-41 University of Copenhagen, Dept. Economics. 2006.

15. Lee, R. B. (1976) *The !Kung San. Men, Women and Work in a Foraging Society.* Cambridge University Press, 1979. Blurton-Jones, N. and Sibly, R. Testing the adaptiveness of culturally determined behaviour: do bushmen women maximize their reproductive success by spacing births widely and foraging seldom? In Reynolds, V. and Blurton-Jones (eds) *Human Behaviour and Adaptation.* Taylor and Francis, London.

16. This is all an area of active research by anthropologists, archeologists and economists. The following references provide an introduction: Cohen, Mark N. 1977. *The Food Crisis In Prehistory.* New Haven: Yale University Press. De Meza, David and J. R. Gould. 1992. The Social Efficiency of Private Decisions to Enforce Property Rights. *Journal of Political Economy* 100: 561-580. Galor, Oded and D. Weil. 2000. Population, Technology, and Growth: From Malthusian Stagnation to the Demographic Transition and Beyond. *American Economic Review* 90: 806-828. Kremer, M. 1993. Population Growth and Technological Change: One Million B. C. to 1990. *Quarterly Journal of Economics* 108: 681-716. Kremer (1993) and Galor and Weil Galor, Oded and D. Weil. 1999. From Malthusian Stagnation to Modern Growth. American Economic Review 89: 150-154. Pryor, Frederic L.

1986. The Adoption of Agriculture: Some Theoretical and Empirical Evidence. *American Anthropologist* 88: 879-897. Pryor, Frederic L. 2004. From Foraging to Farming: The So-Called "Neolithic Revolution," *Research in Economic History* 22: 1-41.

17. Relevant references are Dewar, R. E. (1984). Environmental Productivity, Population Regulation, and Carrying Capacity. *American Anthropologist* 86(3): 601-614. Moore, J. (1983). Carrying capacity, cycles and culture. *Journal of Human Evolution* 12: 505-514. 1983. Zubrow, E. B. W. (1975). *Prehistoric Carrying Capacity: a Model.* Menlo Park, CA, Cummings Publishing Company.

18. Diabetes is a common disorder of glucose metabolism. Insulin, secreted by the pancreas, is instrumental in the uptake of glucose by cells. If the pancreas stops producing insulin (diabetes type 1, usually a genetic predisposition), or if the insulin is insufficient (diabetes type 2, usually caused by overweight and a diet with a high glycemic load), then blood glucose rises above normal, unless the diabetes is treated in some way. Life-threatening diabetes is common today, in some under-developed countries with a high carbohydrate diet, and in many urban communities. For a non-technical introduction to the problems associated with diabetes and diet, see Anthony Worrall Thompson's book *gi diet* Kyle Cathie, London, 2005.

19. Fava beans have an effect upon some genetically susceptible people with severe deficiency of glucose-6-phosphate dehydrogenase (an enzyme). This deficiency, called favism, is thought to confer protection against malaria only in those geographic areas where favism exists. [Golenser 1983]. A substance in fava beans called isouramil triggers the hemolytic anaemia in the deficient individuals, and it is this interaction of isouramil with the enzyme deficient erythrocytes

189

which renders these red blood cells incapable of supporting the growth of the malarial pathogen (*Plasmodium falciparum*). Thus, the spread of agriculture (fava beans in this case) to geographic locations surrounding the Mediterranean was responsible for protecting some people from malaria, but susceptible to anaemia. For further details see Golenser J. *et al.* "Inhibitory effect of a fava bean component on the *in vitro* development of plasmodium falciparum in normal and glucose-6-phosphate dehydrogenase-deficient erythrocytes." *Blood,* vol. 61, pp. 507-510. 1983, and for other deleterious effects of beans see Grant *et al.* The effect of heating on the haemagglutinating activity and nutritional properties of bean (*Phaseolus vulgaris*) seeds." *J Sci Food Agric,* vol. 33, pp. 1324-1326. 1982, and Gupta Y.Antinutritional and toxic factors in food legumes: a review. *Plant Foods for Human Nutrition,* vol. 37, pp. 201-228. 1987.

20.   In mammals, excess food energy is stored as fat in subcutaneous and abdominal depots. The dominant fatty acids in the fat storage depots of wild mammals are saturated fatty acids, whereas the dominant fatty acids in muscle and all other organ tissues are polyunsaturated fatty acids and monounsaturated fatty acids.   Because subcutaneous and abdominal body fat stores are depleted during most of the year in wild animals, most of the total carcass fat is of the poly- and mono-unsaturated type. A year-round dietary intake of high amounts of saturated fatty acids would have not been possible for pre-agricultural humans preying on wild mammals, whereas when a domesticated animal is slaughtered for food, everything is eaten, including the storage fat containing large amounts of saturated fatty acids. For further details see Cordain L, Watkins BA, Florant GL, Kehler M, Rogers L, Li Y. Fatty acid analysis of wild ruminant tissues: evolutionary implications for reducing diet-related chronic disease. *European Journal of Clinical Nutrition,* 56:181-91.2002

21. Denton D. *The hunger for salt. An anthropological, physiological and medical analysis.* New York: Springer, 1984.

22. Fungal diseases in poorly stored grain are a problem even today. Some fungi grow in food grains and produce harmful substances known as mycotoxins. These can cause cancer, liver disorders, and generally weaken the human immune system. According to the FAO, a quarter of the world's food crop output is affected by such toxins each year. They can enter the human food chain when people eat infected grain or consume meat from livestock raised on infected feeds. Mycotoxins do not easily decompose, are not readily broken down in digestion, and even resist cooking or freezing. A recent assessment concluded that 4.5 billion people in the developing world are chronically exposed to uncontrolled amounts of aflatoxin, a type of mycotoxin that occurs in maize, peanuts, sorghum, and some roots and tubers. A single incident of acute aflatoxicosis killed more than 125 Kenyans who ate infected maize in 2004. See online publications by the International Maize and Wheat Improvement Center (CIMMYT), especially their (free) publication: Ortiz, R.,*Food Safety: Ensuring safe, healthy, nutritious food.* 2008.

Chapter 5

1. See McFarland, D. *Problems of Animal Behaviour.* Longman, Harlow, UK 1989, pp 59-86.

2. Harris, M. *Culture, People, Nature.* (4th edn) Harper and Row, New York, 1985, pp 235-236.

3. For scientific evidence see Madson, et. al. Kinship and altruism. A cross-cultural experimental study. *British Journal of Psychology,* 98 (2007), 339-359. For anthropological

evidence, see Sahlins, M. Stone Age Economics. Tavistock Publications, London, 1974, pp 196-204.

4. For reciprocal altruism in animals see Seyfarth, R.M. and Cheney, D.L. Grooming, alliances and reciprocal altruism in vervet monkeys. *Nature*, 308, 541-543, 1984; Trivers, R. The evolution of reciprocal altruism. *Quarterly Review of Biology*, 46, 35-57, 1971; Trivers, R. *Social Evolution*, Benjamin-Cummins, Menlo Park, CA. 1985.

5. Quote is from Woodburn, J. An introduction to Hazda ecology. In Lee, R. and DeVore (eds) *Man the Hunter,* Aldine, 1968. Further discussion of this general point can be found in Sahlins, M. *Stone Age Economics*, Tavistock Publications, 1972 pp 41-99.

6. Sahlins, M. *Stone Age Economics*, Tavistock Publications, 1972. The quote is from p 37. Also Diamond, J. *Guns,Germs and Steel*. Vintage 1998, pp 265-292. The quote is from p 176.

7. See Harris, M. Culture, people, nature. (4th edn) Harper and Row, New York, 1985 for further discussion of the various types of exchange.

8. Yudkin, J. Archaeology and the nutritionist. In Ucko, P.J. & Dimbleby, G.W. (eds) *The domestication and exploitation of plants and animals*. Gerald Duckworth & co., London, 1969.

9.Tannahill, R. *Food in History*. Paladin 1975, p87-88. See also Colquohoun, K. *Taste: The story of Britain through its food.* Bloomsbury Publishing, London, 2008.

10. For an account of early trading practices see Sahlins, M. *Stone Age Economics* Tavistock Publications, 1972. pp 215-275.

11. For a good account of the early trading years of the East India Company see Giles Milton, *Nathaniel's Nutmeg*, Hodder and Stroughton, London, 1999. The quote is from p 74.

12. Tannahill, R. *Food in History.* Paladin 1975, p87-88. See also Colquohoun, K. *Taste: The story of Britain through its food.* Bloomsbury Publishing, London, 2008.

13. The decline in the quality of the food that arrives on the average family table is due to a number of factors. Firstly, the quality of meat from animals that can exercise their muscles is greater that than that of intensively reared animals that have no exercise. Secondly, the stress induced by transporting animals to the slaughter house and of the slaughter process itself has a marked effect on meat quality (see for example M Debut, C Berri, E Baeza, N Sellier, C Arnould, D Guemene, N Jehl, B Boutten, Y Jego, C Beaumont, and E Le Bihan-Duval. Variation of chicken technological meat quality in relation to genotype and preslaughter stress conditions. *Poultry Science*,82, 12, 1829-1838,2003. Thirdly, Irradiation of meat and vegetables, and transport and storage of vegetables result in considerable vitamin loss. See for example Maria.I, Gil, E.N. and Kadar, A.A. Quality Changes and Nutrient Retention in Fresh-Cut versus Whole Fruits during Storage. *J. Agric. Food Chem.*, 54 (12), 4284 -4296, 2006.

14. A good example of the recent evolution of a domestic animal is the turkey. One of the most successful species of modern times is the turkey. Biological success is measured in terms of genetic contribution to future generations. All that matters is the representation of genes in the gene pool. It does not matter how the success is achieved. It can be achieved by honest toil, by sexual attractiveness, by exploiting other

species. The turkey population has grown enormously as a result of genetic changes that have enabled turkeys to exploit humans. Turkeys are now white rather than black. They mature much more quickly than they used, and they are shorter and heavier in the body. The genetic changes that have been responsible for these adaptations have been crucial to the success of the turkey. If white coloration had not been genetically possible, the turkey would not look so attractive when plucked. If shortening of the body had not been genetically possible then a turkey of decent weight would not fit in the normal household oven. Over decades, turkeys have been selectively bred by humans to satisfy market demand. You may think that this example is not valid because it involves humans as a selective agent. But such relationships between species are commonplace in the animal kingdom. For example, many types of ants farm aphids and milk them for their honeydew. Like turkeys, these aphids have undergone genetic adaptations to this role. Like turkeys, the aphids benefit from protection from predators. Some ants build structures to protect their aphids from the weather. They may move them to new pastures when necessary, and like turkeys, the aphid eggs may be especially cared for. Some ants even collect aphid eggs and bring them into their nest for the winter. From the biologist's point of view, there is no difference of principle between the two cases. Both turkeys and aphids have increased their fitness by genetically adapting to the symbiotic pressures of another species. It is true that humans have benefited from the turkey, but from a biological point of view, the turkey is a very successful species. Turkey genes have achieved a huge net rate of increase over many decades. From the human point of view the turkey is a successful product. Turkey sales have shown a huge net rate of increase over many decades. From the biological point of view, the fact that the turkey is farmed by another species is irrelevant. Similar relationships exist throughout the animal kingdom (e.g. ants and aphids). From the human point of view, the fact that the turkey is an animal is irrelevant. It is still a product that can be redesigned by careful selective breeding. Similar relationships have existed between man and other animals for centuries. For an account of the evolution of the chicken from the jungle to the factory farm, see Vissar, M. *Much Depends on Dinner*. Penguin Books, London, 1986, pp 115-154.

15. For an extended discussion see Jared Diamond, Guns,Germs and Steel. Vintage 1998, pp 392-397. Also see Cavalli-Sforza, L.L. *Genes, peoples and Languages*. Penguin Books, 2000.

16. For at least 5000 years travelers have used the Ridgeway, which originally connect the south coast of Britain to the Wash on the east coast. It provided a reliable trading route because the high dry ground made travel easy and provided a measure of protection by giving traders a commanding view, warning against potential attacks. In the Bronze Age the White Horse (at Uffington) was carved into the chalk subsoil, and a stone circle was erected at Avebury. During the Iron Age, inhabitants took advantage of the high ground by building hill forts along the Ridgeway to help defend the trading route. Following the collapse of Roman authority the Ridgeway used as a road for moving armies in response to the Saxon and Viking invasions. In medieval times, the Ridgeway was used by drovers, moving their livestock from the West Country and Wales to markets near London. The Ridgeway still exists today, and was given the status of a National Trail in 1973.

17. Tudge, C. *Feeding People is easy*. Pari Publishing, Grosseto, Italy, 2007.

18. Favism (from the Italian fava, broadbean) is an X-linked recessive hereditary disease, i.e. it normally occurs in males,

but females can be carriers. It is characterized by abnormally low levels of the enzyme (G6PD), which is especially important in red blood cell metabolism. The disorder is characterized by a hemolytic reaction to consumption of broad beans. It is the most common enzyme deficiency disease in the world, affecting approximately 400,000,000 people globally. A side effect of this disease is that it confers protection against malaria, in particular the form caused by Plasmodium falciparum, the most deadly form of malaria. So in certain parts of the world where this form of malaria is prevalent, favism can confer an evolutionary advantage, which is probably why the genetic abnormality persists in certain ethnic groups. Favism has been known since antiquity, but is thought to date back about 10.000 years, when the farming lifestyle created the conditions (i.e. standing water) in which the malaria mosquito can thrive References are Beutler E (1994). "G6PD deficiency". *Blood* 84,11, 3613–36. 1994.and Simoons, F.J. *Plants of Life, Plants of Death*, University of Wisconsin Press, 1999. (see also Chapter 4 note 19).

Chapter 6

1. Demand is the quantity of a good or service that people want to buy. The factors determining demand include the price of the good or service, the price of competing goods, and the customers' incomes. Other factors include weather conditions and the customers' family circumstances. Price elasticity of demand is the proportional increase in quantity of demand divided by the proportional reduction in price. Income elasticity of demand is the proportional increase of quantity demanded at a given price divided by the proportional rise in income.

2. Tudge, C. *Feeding People is easy*. Pari Publishing, Grosseto, Italy, 2007. p 114.

3. Many the ingredients of chicken nuggets sold in the UK do not originate in the UK. Undeclared bovine proteins were found by DNA tests to be present in chicken nuggets from Dutch processors. . Some manufacturers were using the technique of injecting so-called hydrolyzed proteins. These are proteins extracted from old animals or parts of animals which are no use for food, such as bone, feathers, hide, ligaments and skin. They make the flesh swell up and retain liquid. It is unclear what kind of cow products had been used to produce bovine proteins. Certain materials and processes are banned by UK law, in an attempt to control BSE.

Part of the problem is that ingredients for chicken nuggets are sourced from as far away as Brazil and Thailand. Tesco has invested heavily in Thailand and owns a majority share of the retail arm of Thailand's leading chicken producer. KFC is a fast food retailer that sources some of their meat direct from Thailand. Grampian is a supplier of fresh chicken and nuggets to all the main British supermarkets. It has closed down a factory in Scotland and ended its contract with some of its British farmers, and bought two huge factories outside Bangkok. Although Thai poultry imports to the EU are supposed to be checked, some producers are subcontracting to China, or buying in chicken from China and then relabelling it as Thai produce before selling it on, just as British factories buy in poultry from the Netherlands and relabel it as British.

Mechanically separated meat is a paste-like meat, made by forcing animal bones, with attached edible meat, under high pressure through a sieve to separate the bone from the edible meat tissue. It has been used in certain meat and meat products since the late 1960s. Meat slurry is a liquefied meat product that contains fewer fats, pigments and less myoglobin than unprocessed dark meats. It is not designed to sell for general consumption; rather, it is used as a meat supplement in food products for humans, such as chicken nuggets, and food for domestic animals.

4. Clark, S. and Clark, S. *Moro East*, Elbury Publishing, Random House, 2007.

5. Stroke is the rapidly developing loss of brain functions due to a disturbance in the blood vessels supplying blood to the brain. There are many causes of stroke, but a major long-term cause is diet, especially a diet high in sugar, fat, and salt. Such diets tend to lead to high blood pressure and narrowing of blood vessels by fatty deposits. Stroke is the third leading cause of death in the United States, and the number two cause of death world-wide. See online American Heart Association. (2007). *Stroke Risk Factors*; *and Cholesterol, diastolic blood pressure, and stroke.* See also 13,000 strokes in 450,000 people in 45 prospective cohorts. Prospective studies collaboration" (1995). *Lancet* 346 (8991-8992): 1647-53.

Chapter 7

1. Nicklas, T., Fisher, J. 2003. To each his own: family influences on children's food preferences. *Pediatric Basics*. 102:13-16.

2. Bartoshuk, L. M. Psychophysical advances aid the study of genetic variation in taste. *Appetite* 34, 105. 2000. Bartoshuk, L. M., Duffy,V.B. *et al.* (1994). PTC/PROP tasting: anatomy, psychophysics, and sex effects. *Physiol Behav* 56(6): 1165-71. 1994. Bartoshuk, L. M.. Sweetness: History, Preference, and Genetic Variability. *Food Technol.* 45, 108,110, 112-113. 1991. Lanier, SA, JE Hayes, VB Duffy. V.B. (2005). Sweet and bitter tastes of alcoholic beverages mediate alcohol intake in of-age undergraduates. *Physiology & Behavior* 83, 821-831.2005. Drewnowski, A, SA Henderson, S.A. *et al.* Taste

and food preferences as predictors of dietary practices in young women. *Public Health Nutr* 2, 513-9.1999. Drewnowski, A, Henderson,S.A. *et al.* . Genetic taste markers and food preferences. *Drug. Metab. Dispose* .29, 535-8. 2001. Dinehart, ME, Hayes,J.E. *et al.* Bitter taste markers explain variability in vegetable sweetness, bitterness, and intake. *Physiol Behav* 87, 304-13. 2006.

3. For these experiments, those doing A-level biology, recruited pupils from the year below us (age range 13-16), and 18 boys and 15 girls agreed to take part in our first experiment. We prepared sandwiches measuring 1 x 1 x 2 cm, made of white bread and containing coriander leaf or parsley leaf, cooked for precisely 1.5 minutes in sunflower oil. The sandwiches were presented to the subjects in alternating order. Each subject was asked to rate the sandwiches on a 0 to 5 scale of pleasantness. If they chose 0, they were asked to rate the sandwich on a 0 to 5 scale of unpleasantness. The results from anyone who disliked both sandwiches were rejected (4 subjects).

All subjects liked the parsley sandwiches, whereas 9 subjects disliked coriander leaf, and 20 liked it, a ratio of one in 3.2 (chi squared = 0.0367, n = 1). There was no overlap between the scores of those who liked coriander and those who disliked it.
In the second experiment subjects there were 13 girls and 15 boys from the same subject pool as the first experiment. They were presented with pieces of bread (approx. 1 cm$^{3)}$ containing either caraway, or cumin seeds. The testing procedure was the same as in the first experiment.

The results showed that no subjects disliked cumin, whereas 3 girls and 3 boys disliked caraway. So, out of 28 subjects, 6 disliked caraway, a ratio of one in 4.66 (chi squared = 0.86, n =

1). There was no overlap between the scores of those who liked caraway and those who disliked it. Those subjects that disliked coriander were not the same as those that disliked caraway, except in one case.

These results are in accordance with the hypothesis that those that dislike coriander are homozygous for a recessive coriander 'taster' gene, while those that dislike caraway are homozygous for a recessive caraway 'taster' gene.

4. Lucy J Cooke, L.J. Claire MA Haworth, C.M.A. and Wardle, J. Genetic and environmental influences on children's food neophobia. *American Journal of Clinical Nutrition*, 86, 428-433, 2007

5. Burghardt, G. and Hess, E. Food Imprinting in the Snapping Turtle, *Chelydra serpentina Science*, 151. 3706, 108 – 109, 1966: Grassman, M. Chemosensory Orientation Behavior in Juvenile Sea Turtles *Brain Behav Evol* 1993;41:224-228: Gordon M. Darmaillacq, A. Raymond Chichery, R. and Ludovic Dickel, L. Food imprinting, new evidence from the cuttlefish *Sepia officinalis Biol Lett.* 2006 September 22; 2(3): 345–347. Published online 2006 April 25. doi: 10.1098/rsbl.2006.0477.

6. Wardle,J., Guthrie,C., Sanderson,S., Birch, L., and Plomin,R., Food and activity preferences in children of lean and obese parents *International Journal of Obesity,* 25, 971-977, 2001; Faith MS, Rha SS, Neale MC, and Allison DB. Evidence for genetic influence on human energy intake: results from a twin study using measured observations. *Behaviour Genetics* 29: 145 1999; De Castro JM. Behavioral genetics of food intake regulation in free-living humans. *Nutrition* 15: 550, 1999. De Castro JM. Heritability of hunger relationships with food intake in free-living humans. *Physiol Behav* 1999; 67:

249, 1999: Barker, D. J. P. Mothers, Babies and Disease in Later Life. *London: BMJ Publishing Group*, 1994. Summarizes much of the data leading to development of the fetal origins hypothesis.

Waterland, R. A., and Cutberto, G. Potential Mechanisms of Metabolic Imprinting That Lead to Chronic Disease. The *American Journal of Clinical Nutrition* 69 (1999) Reviews data in support of metabolic imprinting and discusses potential underlying biological mechanisms.

7. Brunstrom, J.M. Dietary learning in humans: Directions for future research. *Physiology & Behavior*, 85, 57-65, 2005. Brunstrom, J.M.and Mitchell, G.L., Flavor–nutrient learning in restrained and unrestrained eaters. *Physiology & Behavior*, 90, 133-141, 2007 Brunstrom, J.M.. Associative learning and the control of human dietary behavior. A*ppetite*, 49, 268-271, 2007

8. Sugar consumption per head increased 30 fold between the mid 18th century and the mid 20th century. This was made possible by establishing vast sugar plantations in various parts of the world (often with slave labour). The increase in the popularity of sugary foods was almost certainly due to a human open-ended preference for sweet tasting food.

Yudkin, J. Archaeology and the nutritionist. In Ucko, P.J. & Dimbleby, G.W. (eds) *The domestication and exploitation of plants and animals.* Gerald Duckworth & co., London, 1969. Quotes are from p 551 and 552. See also Yudkin, J. Evolutionary and historical changes in dietary carbohydrates. *Am. J. med. Ass.,* 177, 316, 1967. Yudkin, J. Sugar and Coronary Thrombosis, *New Scientist,* March 16, 1967. Yudkin, J., *Pure, white and Deadly*, Penguin. 1972 and 1988

9 For an account of the vicissitudes of salad see Visser, M. *Much Depends on Dinner.*Penguin 1986, pp 192-223. The ice cream quote is from p 309-310. For an up to date account of food additives see Wilson, B. *Swindled*, John Murray, 2008.

10 In Chapter 1 we saw how the liver can become conditioned to reduce the rate of conversion of glycogen to glucose just before a meal, in anticipation of the glucose load that will arrive as a result of digesting the meal. The liver can do this only if the meals are at roughly the same time each day. In other words, the liver plays its part in maintaining the physiological harmony of the body, by taking action at the right time. The conversion of glycogen to glucose is only one of the many metabolic processes that are attuned to our biological rhythms. Nearly all animals have such rhythms (an indication of their fundamental importance); they influence our sleeping and waking, and our feeding and drinking.

11. Sulfites may be added to food as an enhancer and preservative. About one out of 100 people is sensitive to sulfites. Such sensitivity can develop at any time in life, and the cause of sensitivity is unknown. Allergic reactions to sulfites can be mild or life threatening. In 1986, the Food and Drug Administration (USA) banned the use of sulfites on fruits and vegetables that are eaten raw, such as lettuce or apples. For a general account of the additive problem, see Lawrence, F, *Additives: A Survival Guide*. Edbury Press, 1986.

Chapter 8

1. See Sahlins, M. *Stone Age Economics* , especially pp 9-14 and 246-263.

2.For a general discourse on social status see Bourdieu, P. *Distinction: a Social Critique of the Judgment of Taste,* translated by Richard Nice. Cambridge: Harvard University Press, 1984. Marmot, M. *The Status Syndrome: How social status affects our health and longevity.* Holt Paperbacks, 2005.

3. Britton, L.E., Martz, D.M., Bazzini,D.G., Curtin,L.A. and Shomb, A.L. Fat talk and self-preservation of body image: Is there a social norm for women to self-degrade? *Body Image,* 3, 247-254, 2006. Tucker, K.L., Martz, D.M., Curtin, L.A. and Bazzini, D.G. Examining "fat talk" experimentally in female dyad: How are women influenced by another woman's body presentation style? *Body Image,* 4, 157-164, 2007.

4. Klump KL, Kaye WH, Strober M (2001) The evolving genetic foundations of eating disorders. *Psychiatr Clin North Am,* 24 (2), 215-25. Wade TD, Bulik CM, Neale M, Kendler KS. (2000) Anorexia nervosa and major depression: shared genetic and environmental risk factors. *Am J Psychiatry,* 157 (3), 469-71.

Lindberg L, Hjern A. (2003) Risk factors for anorexia nervosa: a national cohort study. *International Journal of Eating Disorders,* 34 (4), 397-408.

5. For the full Kellog story see Visser, M. *Much Depends on Dinner,* Penguin Books, 1986, pp 22-55

6. Lin BH, Guthrie J and Frazao E Nutrient contribution of food away from home. In: Frazao E (Ed). America's Eating Habits: Changes and Consequences. *Agriculture Information Bulletin* No. 750, US Department of Agriculture, Economic Research Service, Washington, DC, pp. 213–239. 1999. Keith SW, Redden DT, Katzmarzyk PT, *et al.* Putative contributors to the secular increase in obesity: exploring the roads less

traveled. *Int J Obes* (Lond) 30 (11): 1585-94. 2006. Farooqi S, O'Rahilly S Genetics of obesity in humans. *Endocr. Rev.* 27 (7): 710–18. 2006. Wardle J, Carnell S, Haworth CM, Plomin R Evidence for a strong genetic influence on childhood adiposity despite the force of the obesogenic environment. *Am. J. Clin. Nutr.* 87 (2): 398-404. 2008. Frayling TM, Timpson NJ, Weedon MN, et al A common variant in the FTO gene is associated with body mass index and predisposes to childhood and adult obesity. *Science* 316 (5826): 889-94. 2007 Chakravarthy MV, Booth FW .Eating, exercise, and "thrifty" genotypes: connecting the dots toward an evolutionary understanding of modern chronic diseases. *J. Appl. Physiol.* 96 (1): 3-10. 2004

Bray, G.A. Medical consequences of obesity . *J. Clin. Endocrinol. Metab.* 89, 2583-9, 2004.

7. Bulimia nervosa is an eating disorder involving episodes of overeating, followed by rapidly removing food from the body before it can be digested. The food is removed either by vomiting (by triggering the gag reflex, or by ingesting emetics), or by using laxatives or enemas. References are Walsh, B T (1995), Pharmacotherapy of eating disorders, *Eating Disorders and Obesity: A Comprehensive Handbook* (New York: Guilford): pp. 329-340, 1995: Palmer, Robert, Bulimia nervosa: 25 years on. *British Journal of Psychiatry,* 185, 447-448. 2004.

Chapter 9

1. Cholesterol is essential for life, and it is synthesized de novo within the body. Cholesterol is recycled within the body. It is excreted by the liver via the bile into the digestive tract. Typically about 50% of the excreted cholesterol is reabsorbed by the small bowel back into the blood stream.

Dietary sources of cholesterol include cheese, beef, pork,

poultry, shrimp and human breast milk. Total fat intake plays a larger role in blood cholesterol than intake of cholesterol itself. Saturated fat is present in full fat dairy products, animal fats, several types of oil and chocolate. Trans fats may be derived from the partial hydrogenation of unsaturated fats, and in contrast to other types of fat, they are not essential for life. It is recommended that trans fats be consumed extremely rarely or not at all as they are said to be more harmful than naturally occurring oils. Trans fat can be found in the commercial food supply including fast food, snack foods, fried food and baked goods.

The view that a change in diet (specifically, a reduction in dietary fat and cholesterol) can lower blood cholesterol levels, and thus reduce the likelihood of development of, amongst others, coronary artery disease, has been challenged. An alternative view is that any reductions to dietary cholesterol intake are counteracted by the organs such as the liver, which will increase or decrease production of cholesterol to keep blood cholesterol levels constant. See Olson,R.E. Discovery of the lipoproteins in fat transport and their significance as a risk factor. *J. Nutrition*, 128, 439S-443S, 1998. Lewington S, Whitlock G, Clarke R, Sherliker P, Emberson J, Halsey J, Qizilbash N, Peto R, Collins R.. "Blood cholesterol and vascular mortality by age, sex, and blood pressure: a meta-analysis of individual data from 61 prospective studies with 55,000 vascular deaths". *Lancet* 370 (9602): 1829–39. 2007. Uffe Ravnskov *The Cholesterol Myths : Exposing the Fallacy that Saturated Fat and Cholesterol Cause Heart Disease*. New Trends Publishing, Inc. 2000:Daniel Steinberg. *The Cholesterol Wars: The Cholesterol Skeptics vs the Preponderance of Evidence*. Boston: Academic Press. 2007

2. For a detailed account of the war between margarine and butter producers see Visser, M. *Much Depends on Dinner*, Penguin Books, 1986, pp 83-114.

3.References are: McBean, L. D. and Speckmann, E.W. Food faddism: a challenge to nutritionists and dietitians. *American Journal of Clinical Nutrition*, 27, 1071-1078. 1974: Roberts, I.F., West, R.J., Ogilvie, D, and Dillon, M.J. (1979). Malnutrition in infants receiving cult diets: a form of child abuse. *British Medical Journal 1*(6159): 296–298. 1979: Carey,

S (2004). *A beginner's guide to the scientific method*. Third Edition. Belmont, CA: Wadsworth/Thomson Learning. 2004: Katz, D.L., Pandemic obesity and the contagion of nutritional nonsense. *Public Health Reviews* 31, 33-44. 2003

4. Oil derived from the tissues of oily fish is recommended for a healthy diet, because it contains omegea-3 fatty acids. The fish obtain these fatty acids by consuming certain microalgae, or by preying upon other fish that have eaten the microalgae. However, species at the top of the food chain may accumulate certain toxins. For this reason, the authorities (e.g. the FDA) recommend limiting consumption of certain predatory fish species, such as Tuna, that may have high levels of mecury, dioxin, PCBs or chlordane. References include EPA *Fish consumption advisories*,2007. *Fish and Omega-3 Fatty Acids*. American Heart Association, 2007.

5. Rena R Wing and Suzanne Phelan. Long-term weight loss maintenance. *American Journal of Clinical Nutrition*.2005. L. Stahre et al., A short-term cognitive group treatment program gives substantial weight reduction up to 18 months from the end of treatment. A randomized controlled trial. *Eating and Weight Disorders*. 10. p 51-58 2005. American Dietetic

Association. Position paper on vegetarian diets. *J Am Diet Assoc.* 103:748-765. 2003. Davis, B. and Melina, V. 2000. *Becoming Vegan.* p. 22. 2000. Wansink, B. *Mindless Eating. Why we eat more than we think.* New York: Bantam Dell, 2006.

6. According to the principles of thermoregulation, humans are endotherms. We expend energy to maintain our body temperature at about 37 °C (98.6 °F). This is accomplished by metabolism and blood circulation, by shivering to stay warm, and by sweating to stay cool. In addition to thermoregulation, humans expend energy keeping the vital organs (especially the lungs, heart and brain) functioning. Except when sleeping, our skeletal muscles are working, typically to maintain upright posture. The average work done just to stay alive is the basal metabolic rate, which (for humans) is about 1 watt per kilogram of body mass (0.45 W/lb). Thus, an average man of 75 kilograms (165 lb) who just rests (or only walks a few steps) burns about 75 watts per day, or 1 kilocalorie each minute.

7. Acomplia (rimonabant) is an anti-obesity drug. It was approved for marketing in the European Union in June 2006. It is not yet approved for use in the United States, where it is known as Zimulti The drug acts by selectively blocking CB1 receptors found in the brain and in peripheral organs important in glucose and lipid metabolism, including adipose tissue, the liver, gastrointestinal tract and muscle. It inhibits those brain circuits that make people hungry when they smoke cannabis. It acts to decrease the overactivity of the endocannabinoid system. This is a recently characterised physiological system that includes receptors such as the CB1 receptor and it has been shown to play an important role in regulating body weight and in controlling energy balance, as well as glucose and lipid (or fat) metabolism. In clinical studies, it has been shown to

improve a wide array of cardiometabolic risk factors as well as promoting sustained weight loss. References are Pagotto U. Pasquali R. Fighting obesity and associated risk factors by antagonising cannabinoid type 1 receptors. *Lancet.* 2005; 365: 1363-64.Van Gaal LF, Rissanen, AM, Scheen AJ, Ziegler O, Rössner S for the RIO-Europe Study Group. Effects of the cannabinoid-1 receptor blocker rimonabant on weight reduction and cardiovascular risk factors in overweight patients: 1-year experience from the RIO-Europe study. *Lancet.* 2005; 365: 1389-97.Marzo V, et al. Leptin-regulated endocannabinoids are involved in maintaining food intake. *Nature.* 2001,410, 822-825.Després, J.P. et al. Effect of Rimonabant on Body Weight and the Metabolic Syndrome in Overweight Patients, *New England Journal of Medicine,* November, 16, 2005. Van Gaal L, et al. Effects of The Cannabinoid-1 Receptor Blocker Rimonabant On Weight Reduction And Cardiovascular Risk Factors In Overweight Patients: 1-Year Experience From The RIO-Europe Study. *The Lancet.* 365;1389-1397.

8. Cabanac, M. Palatability versus money: Experimental study of a conflict of motivations. *Appetite,* 25, 43-49, 1995.

9. Rolls. B. *Volumetrics eating plan.* Harper Collins, 2005. The Volumetrics Eating Plan focuses on the scientific principles of satiety, by enhancing the feeling of fullness while simultaneously consuming fewer calories. Foods with higher water content have fewer calories but control appetite better, i.e., fruits, vegetables and soups are 80-95 % water whereas oils are 100% fat, 0% water. Rolls' studies on hunger/satiety reveal that most people eat the same weight or volume of foods at meals. Hence, by eating nutritious foods with lower caloric value, dieters can experience a feeling of fullness and improve nutrient intake while losing weight.

10. Kathleen DesMaisons, K. (2000). *The Sugar Addict's Total Recovery Program*. Ballantine Books. 2000. Carlo Colantuoni, c., Rada, P., McCarthy, J., Patten, C.,. Avena,N.M., Chadeayne, A. and Hoebel, B.G. Evidence That Intermittent, Excessive Sugar Intake Causes Endogenous Opioid Dependence. *Obesity Research* 10,478-488, 2002.Nurnberger, Jr., J I., and Bierut, L. J.Seeking the Connections:Alcoholism and our genes. *Scientific American,* 296, 2007.Dick DM and Bierut LJ, The Genetics of Alcohol Dependency, *Current Psychiatric Reports*, 8, 151-7, 2006.

11. Schlosser, E. *Fast Food Nation. The dark side of the all-American meal.* Harper Perrennial, 2005.

12. Quackwatch is a web-based guide to quackery, health fraud and intelligent decisions. It is operated by Dr. Stephen Barrett. www.quackwatch.org

13. Jacobus,C.H., Holick,M.F., Shao,Q, Chen,T.C., Holm,I.A., Kolodny,J.M., Fuleihan, G.E., and Seely E.W. Hypervitaminosis D associated with drinking milk. *New England Journal of Medicine,* 326, 1173-1177, 1992

Chapter 10

1.David Beresford. *Ten Men Dead*, Atlantic Press, New York, 1987

2. Self-destructive behaviour performed for the benefit of others is called altruism by biologists. This usage contains no implications about intentions or motives. In genetic terms, genes for parental care tend to be preserved in the bodies of the surviving offspring, and such genes are, therefore, likely to increase in frequency relative to the genes that promote neglect towards offspring. Thus altruism at the individual level is a

manifestation of 'selfishness' at the gene level. In other words, Altruism towards kin can be regarded as selfishness on the part of the genes responsible, because copies of these genes are likely to be present in relatives.

3. Dawkins, R. *The extended phenotype*. W.H. Freeman, Oxford. 1982. A phenotype is the bodily expression of genetic influence. In this book, Dawkins argues that genes can have extended phenotypes outside the bodies in which they sit. Examples are the characteristic nests of particular species of birds, the species-

distinctive houses of caddis flies, and anything that is the result of the typical behaviour of a species. So tools, such as bows and arrows, of hammers and chisels, are part of the human extended phenotype, as are the items achieved by the use of those tools.

4. Cultural behavior is behaviour that has been passed from one generation to another by non-genetic means. Evolution occurs primarily as a result of natural selection, and direct genetic inheritance of acquired characteristics is not possible. However, some genes can be influenced environmental events that are themselves the result of culture (e.g. diet) (see Laland, K.N, , Odling-Smee, J. and Myles, S. 2010 How culture shaped the human genome: Bringing genetics and the human sciences together. *Nature Reviews Genetics* 11: 137-148, 2010.). Moreover, information can be passed from parent to offspring through the processes of imprinting and imitation. Sensitive periods of learning occur in the early life of many animals, and during such periods they often learn from their parents. Some song -birds remember the parental song, provided they hear it during the sensitive period (about 10 to 50 days old), and are able to sing in later life. The tendency to copy the parental song leads to regional variation. Populations separated by only a few

miles may have different dialects. Animal dialects represent an elementary form of tradition. Other forms of traditional behaviour include migration routes and feeding habits.

5. Both innate and adaptive immunity depend on the ability of the immune system to distinguish between self and non-self molecules. The non-self molecules are recognized by the immune system as foreign invaders. Some of these are antibody generators (antigens), and they bind to specific immune receptors and elicit an immune response. In other words they are attacked by the immune system and rendered inactive. References are Beck, G., Gail S. Habicht, G.S. Immunity and the Invertebrates. *Scientific American*: 60–66. 2007. Litman G, Cannon J, Dishaw L Reconstructing immune phylogeny: new perspectives." *Nat Rev Immunol.* 5 (11): 866-79. 2005. Smith A.D. (Ed) *Oxford dictionary of biochemistry and molecular biology.* (1997) Oxford University Press.

6, Disorders in the immune system can occur when the immune system is less active than normal, resulting in recurring and life-threatening infections. These immunodeficiency diseases can either be the result of a genetic disorder, or be produced by an infection, or by pharmaceuticals. Sometimes a hyperactive immune system attacks normal tissues as if they were foreign invaders. Such hypersensitivity is an immune response that damages the body's own tissues. One type of hypersensitivity (called type 1) produces allergic reactions. These are commonly associated with asthma, eczema, food allergies, hat fever and the venom of bees and wasps. In some people, severe allergies to environmental or dietary allergens (or to medication) may result in life-threatening anaphylactic reactions and potentially death.

7. A food allergy is an exaggerated immune response triggered

by specific foods such as milk, eggs, and peanuts. Normally, the body's immune system defends against potentially harmful substances, such as bacteria, viruses, and toxins. In some people, an immune response is triggered by a substance that is generally harmless, such as a specific food. In a true food allergy, as opposed to other forms of food intolerance, the immune system produces antibodies and histamine in response to the specific food. Any food can cause an allergic reaction, but certain foods common culprits. These include milk, eggs, peanuts, shellfish, nuts, and wheat. Food allergy is on the increase in some countries, and this may be due to food additives, such as such as dyes, preservatives, and thickeners. A food allergy frequently starts in childhood, but it can begin at any age. Fortunately, many children will outgrow their allergy to milk, egg, wheat, and soy by the time they are 5 years old if they avoid the offending foods when they are young. Allergies to peanuts, tree nuts, and shellfish tend to be lifelong. References are American Gastroenterological Association medical position statement: guidelines for the evaluation of food allergies. *Gastroenterology*.120,1023-5. 2001. American College of Allergy, Asthma, & Immunology. Food allergy: a practice parameter. Ann *Allergy Asthma Immunol*. 96(3 Suppl 2):S1-68. 2006. Adkinson N.F Jr. *Middleton's Allergy: Principles and Practice*. 6th ed. Philadelphia, Pa: Mosby; 2003.

8. Gluten intolerance, also called celiac disease, is an autoimmune disorder of the small intestine that occurs in genetically predisposed people of all ages from middle infancy. Celiac disease is caused by a reaction to a gluten protein found in wheat, barley and rye. The disease affects the absorption of nutrients and, the only effective treatment is a lifelong gluten-free diet. Note that, while the disease is caused by a reaction to wheat proteins, it is not the same as wheat allergy.

9. The activities of the food corporates have received much critical attention recently. See Pollan, M. *In Defence of Food: The Myth of Nutrition and the Pleasures of Eating*. Allen Lane, 2008, and Patel, R. *Stuffed and Starved: Markets, Power and the Hidden Battle for the World Food System*. Portobello, 2008. Lawrence, F. *Not on the Label: What really goes into the food on your plate*. Penguin, 2004, and Lawrence, F. *Eat your Heart out: Why the food business is bad for the planet and your health*. Penguin, 2008.

10. Tudge, C. *Feeding People is Easy* . Pari Publishing, Grosseto, Italy, 2007, The quote is from p 7.